Proceedings of the Institution of Mechanical Engineers

I Mech E

Engineering Solutions to the Management of Solid Radioactive Waste

International Conference

12–14 November 1991
Ramada Renaissance Hotel, Manchester

Sponsored by
Nuclear Energy Committee of the Power Industries Division of the
Institution of Mechanical Engineers

In association with
British Nuclear Forum
Institution of Nuclear Engineers
Institution of Civil Engineers

IMechE 1991–14

Published for IMechE by
Mechanical Engineering Publications Limited

First Published 1991
This publication is copyright under the Berne Convention and the
International Copyright Convention. Apart from any fair dealing for the
purpose of private study, research, criticism or review, as permitted
under the Copyright, Designs and Patents Act, 1988, no part may be
reproduced, stored in a retrieval system, or transmitted in any form or by
any means, electronic, electrical, chemical, mechanical, photocopying,
recording or otherwise, without the prior permission of the copyright
owners. Reprographic reproduction is permitted only in accordance with
the terms of licences issued by the Copyright Licensing Agency,
90 Tottenham Court Road, London W1P 9HE. *Unlicensed multiple
copying of the contents of this publication is illegal.* Inquiries should be
addressed to: The Managing Editor, Mechanical Engineering Publications
Limited, Northgate Avenue, Bury St. Edmunds, Suffolk, IP32 6BW.

Authorization to photocopy items for personal or internal use, is granted
by the Institution of Mechanical Engineers for libraries and other users
registered with the Copyright Clearance Center (CCC), provided that the
fee of $0.50 per page is paid direct to CCC, 21 Congress Street, Salem,
Ma 01970, USA. This authorization does not extend to other kinds of
copying such as copying for general distribution for advertising or
promotional purposes, for creating new collective works, or for resale,
085298 $0.00 + .50.

© The Institution of Mechanical Engineers 1991

ISBN 0 85298 768 4

A CIP catalogue record for this book is available from the British Library.

The Publishers are not responsible for any statement made in this
publication. Data, discussion and conclusions developed by authors are
for information only and are not intended for use without independent
substantiating investigation on the part of potential users.

Printed by Waveney Print Services Ltd, Beccles, Suffolk

Contents

Transport and Packaging of Radioactive Waste
Engineered Solutions from Croft Associates

As a company our sole business interest is in the design, licensing and supply of packaging for the transport of a wide range of radioactive materials. We have one of the largest teams of people in the UK dedicated totally to this activity.

We supply a wide range of packaging for the transport of radioactive materials. In particular, as the main contractor to all the major nuclear organisations in the UK for LLW packaging, we have supplied to date in excess of 1200 Strong Industrial and IP-2 freight containers for the transport and disposal of radioactive waste.

We can advise on all aspects of packaging and transport, for solid and liquid radioactive materials, including IP, Type A and Type B package designs for fissile and non-fissile materials. Advice is also available on the implementation of the 1985 IAEA Regulations introduced in the UK by recent legislation.

If you have a requirement for packaging and transport for radioactive materials contact Croft Associates for professional and expert advice.

Croft Associates Limited
Upton Lodge, Upton, Oxfordshire,
OX11 9HR, United Kingdom

Telephone: 0235 851151, Fax: 0235 851288

C431/072

Commercial low level waste disposal sites in the United States: the past and the future

F W GARDNER, BS, MS and R T ANDERSON, MS
Chem-Nuclear Environmental Services Incorporated, Columbia, South Carolina,
United States of America

SYNOPSIS

The design and operations of the Barnwell Low-Level Radioactive Waste (LLR) Disposal Facility is compared to new "Triple-Safe" designs being developed for new LLRW sites in the U.S. The new sites will be designed to provide multiple, engineered barriers and receive a much smaller volume of waste, as compared to Barnwell.

The new sites performance requirements will accommodate new regulations, concerns of the public regarding operational and post-operational monitoring and unique applications of composite material construction.

These new features will be used to obtain public acceptance, regulatory approval, and provide long-term stability of the wastes.

INTRODUCTION

The Barnwell site received almost seventy percent of the U.S. LLRW waste from 1971 to 1991. Design and operations of the Barnwell facility are compared to future sites to be constructed and operated in the U.S. Future site design, construction, and operations are discussed for new sites in North Carolina, Illinois, Pennsylvania and Connecticut. The adaptability of the new site design--the triple safe concept--and the flexibility of this design to meet variations in performance requirements and economics are highlighted. This paper concludes that the triple safe design, based on a proven track record of site construction and operations experience, can be adapted to a wide range of operating and performance requirements.

The federal government and the individual states designated as hosts for the disposal of Low-Level Radioactive Waste (LLRW) have made the requirements for disposal more stringent than in the past. As a result, advanced technologies employing multiple engineered barriers are being developed to supplant the existing practice of shallow land burial.

The Barnwell Waste Management Facility was developed under licensing and regulatory requirements in effect in the late 1960s. During the 1970s new regulations were developed by the US Nuclear Regulatory Commission to further clarify the requirements for licensing, design and operation of low level waste burial facilities in the United States. These new regulations are codified in the U.S. Code of Federal Regulations: Title 10 Part 61--also called 10CFR61.

The development, design and operation of Barnwell Waste Management Facility is recognized as a model LLRW disposal facility. It holds the distinction of being the only commercial U.S. disposal facility that has never been closed by regulatory action. In the environmental assessment of

the Barnwell facility, the U.S. Nuclear Regulatory Commission (1982) concluded that the facility had no significant impact on the environment, and that "the Barnwell facility is operated at the forefront of technology related to the shallow land burial of low-level waste."

SITE DEVELOPMENT HISTORY

The development of the Barnwell facility began in 1969 when Chem-Nuclear Systems, Inc. purchased land about 5 miles (8 km) west of Barnwell, South Carolina. The burial facility consists of about 300 acres (121 ha), of which 235 acres (95 ha) have been deeded to the State of South Carolina. This disposal area, for which Chem-Nuclear holds a long-term lease, is used for disposal operations. Surrounding the site Chem-Nuclear also owns over 1000 areas of land--sometimes referred to as a buffer zone. In addition to disposal trenches, the site contains administration, transportation, maintenance, decontamination, and support facilities.

KEY SITE SELECTION CRITERIA

At the time the Barnwell site was developed, the primary defense against migration of radioactive material into the biosphere was the geological containment provided by the sub-surface structure of the land and carefully engineered shallow land burial techniques. The site is located in a coastal plain area receiving 63 inches of rain annually.

Because water management is of particular concern in this humid environment, the following geohydrological parameters were evaluated during pre-construction geohydrological investigations:

- The extent and type of earth materials through which hazardous substances may migrate;
- Ion-exchange adsorption characteristics of the sub-surface materials;
- Vertical and horizontal permeability characteristics of these materials and the corresponding hydraulic radiance likely to be developed;
- Location of the site within the regional ground and surface water flow system;
- The availability, quality, and uses (present and potential) of ground water and surface water in the area.

BASIC DESIGN CRITERIA-LLRW SITES

New technologies employed rely on multiple engineered barriers to isolate wastes. Chem-Nuclear has developed site specific criteria based on USNRC criteria in 10CFR61, state regulations and public input. Each site is subject to unique criteria. Generally, Chem-Nuclear design criteria and goals are summarized in Table 1.

Table 1 - New LLRW site Design Criteria/Goals

- Environmental Dose - To maximum exposed individual outside facility - 25 mrem/yr. Target Goal - 1mrem/yr. (ALPHA)
- Worker Dose - To occupationally exposed personnel inside facility - 5000 mrem/yr. Target Goal - Average 250 mrem/yr.
- Intruder Dose Limit - Maximum individual dose within 300-500 years past closure - maximum 50 mrem/yr.
- Stability - 300-500 years.
- Monitoring
- Receivability

VOLUME OF WASTE

Operations of the Barnwell site began in 1971, and by 1978 the site was receiving over 80 percent of the nation's low-level radioactive waste for burial. In 1981, the State of South Carolina imposed volume limitations on the site to limit the amount of radioactive material buried. Currently, up to 100 000 cubic feet (2832 m³) of radioactive material may be buried each month. New sites are expected to receive substantially smaller waste volumes. Table 2 shows the projected volumes of wastes compared to Barnwell.

LICENSING

Chem-Nuclear Systems, Inc. currently operates the Barnwell Low-Level Waste Disposal Facility under two licenses: (1) Radioactive Material License #097 issued by the South Carolina Department of Health and Environmental Control (DHEC), and (2) Radioactive Materials License #46-13536-01 issued by the United States Nuclear Regulatory Commission (NRC). The NRC license specifically addresses the possession of special nuclear materials. No permits for hazardous waste are in effect. At new LLRW sites similar licenses will be held for operations.

Table 2 Evolution of LLRW Disposal

	Past	Present	Future
Technology	Simple Landfill	Advanced Landfill	Multiple Engineered Barrier
Position	Below Ground	Below Ground	Above Ground
Waste Receipt Rate	1-2 Million ft³/year	1-2 Million ft³/year	50-500 Thousand ft³/year
Handling	"Kick & Roll --Open Air	"Crane-- Open Air	Crane-- Enclosed
Record Keeping	Simple	Detailed/ Computer	On-Line Computer
Containers	Wood Boxes 55-G/Drums	Metal/ HICs	Metal/ HICs Encapsulate

PACKAGING RESTRICTIONS

Restrictions placed on waste forms help ensure the long term safety of the LLRW sites. All radioactive wastes received for burial at the site must be packaged in containers that meet U.S. Nuclear Regulatory Commission and U.S. Department of Transportation requirements. Cardboard, corrugated paper, or fiberboard drums and boxes are not acceptable containers.

No liquid wastes or solid wastes with free-standing water can be received at the Barnwell site except in accordance with Chem-Nuclear's South Carolina Radioactive Materials License #97. Liquids must be solidified in an NRC/South Carolina DHEC approved medium.

High concentration wastes (1 µCi/cc or greater of isotopes with half-lives of 5 years or more) must have additional containment, which can be provided using: (1) solidification or (2) High Integrity Containers (HICs). HICs, generally constructed of cross-linked polyethylene, or an austenitic steel alloy, are designed to last at least 300 years and withstand the overburden loads experienced in the disposal trenches.

© IMechE 1991 C431/072

Oil or organic solutions, solidified or otherwise, are not acceptable for disposal at Barnwell. The waste generator must certify that its radwaste packages contain less than one percent by volume of unintentional oil. Hazardous wastes are not accepted routinely at the site.

Additional restrictions apply to chelating agents, transuranic elements, special nuclear material (fissile material), pyrophoric materials, biological materials, gaseous wastes, sharp objects, and shipments potentially posing unusual hazards. At new compact sites the wastes will be placed into concrete overpack containers prior to placement into the disposal vault, thus adding an additional layer of packaging.

The process in-use for Class B and C wastes at Barnwell provides the first of three barriers to radionuclide migration in the new "Triple Safe Disposal Site Design".

DESCRIPTION OF BURIAL OPERATIONS

Prior to any shipment of waste, a completed South Carolina Prior Notification Certification Form must be received by the South Carolina DHEC and Chem-Nuclear 72 hours prior to the shipment entering South Carolina. Additionally, shippers must notify Chem-Nuclear by telephone when a shipment departs their facilities.

Fifteen to twenty shipments of radioactive waste usually arrive at the Barnwell disposal site every day. Upon arrival at the site, all shipments are surveyed for contamination and radiation levels, and shipping papers are checked for accuracy and completeness. Waste packaging is checked for compliance with burial site criteria, state and federal license requirements, and Department of Transportation specifications. In addition to Chem-Nuclear's inspections, an inspector from the South Carolina Department of Health and Environmental Control checks all shipments to monitor Chem-Nuclear's performance and the shipper's compliance with regulations. Periodically, incoming shipments are inspected by the U.S. Nuclear Regulatory Commission and Department of Transportation auditors. After inspectors have cleared a shipment for disposal, the truck is directed to the burial trench for off-loading.

STABILITY OF WASTES

Currently Chem-Nuclear is operating two standard trenches: an "unstabilized waste" trench and a "stabilized waste" trench. "Stable" waste has been solidified or otherwise stabilized with an approved medium that has been evaluated in accordance with the stability guidelines outlined in the U.S. NRC Branch Technical Position on Waste Form. According to the waste classification system of 10CFR61, Class A wastes do not need to be stabilized because of their low activity.

All Class B and C wastes must be stabilized in a form that will maintain its integrity for at least 300 years. Class A wastes may be stabilized if the generator so desires.

In 1988, Chem-Nuclear began placing high integrity containers and dewatered ion exchange resins--Class B and C wastes in unstabilized form--into concrete overpacks to meet the stability requirements for Class B and C wastes prior to burial. This process is similar to the overpacking that will be done at the new sites. In this operation, burial containers, also called cask liners, are removed from the shipping and transportation casks and placed directly into concrete overpacks in the burial trench. These concrete overpacks are then sealed by installation of a large lid. These concrete overpacks prevent trench subsidence and provide structural stability to the Class B and C wastes. Use of concrete overpacks also reduces worker exposures and costs associated with the processing of waste prior to transportation and disposal.

INTRUDER BARRIERS

In addition to the stabilization requirements of 10 CFR 61, Class C wastes must be protected against inadvertent intrusion. This is accomplished by placing a concrete intruder barrier over the trench or by burying the wastes at least 16.4 feet (5 m) below the ground's surface. At Barnwell, Class C wastes are placed at the bottom of the trench, thus meeting the depth requirement for inadvertent intrusion. All wastes buried in the slit trench are covered with an intruder barrier.

TRENCH ENGINEERING, CONSTRUCTION AND SITE STABILIZATION

Standard Class A trenches are used for the majority of low-level waste disposal. Standard Class A trench dimensions are generally 100 feet (30 m) wide by 1000 feet (305 m) long and 22 feet (6.7 m) deep. The width of "stabilized" waste trenches is normally 28 feet (8.5 m). The trench floor slopes gently to one side and to one end to collect any infiltrated water for sampling and removal. Illustration One, Typical LLRW Burial Trench shows a typical burial trench used at Barnwell.

In standard trenches, a two-foot (0.6-m) French drain is constructed along the floor's lower side and filled with small stone. Monitoring pipes are installed at intervals not exceeding 125 feet (38 m) to allow for future sampling. Pipes are capped at the top and bottom; the bottom cap has a 0.25-inch (0.6-cm) hole in it. Two sumps, 4 feet x 4 feet (1.2 meters x 1.2 m) and extending 4 feet (1.2 m) below the trench floor, are constructed at 500-foot (152-m) intervals and filled with stone. The sumps provide a place where water can collect and be removed. Up to three feet (0.9 m) of sand is added to the trench floor to provide a firm and level base for the waste and to provide a porous medium for any rainfall that falls in the open trench to move easily to the French drain and sumps.

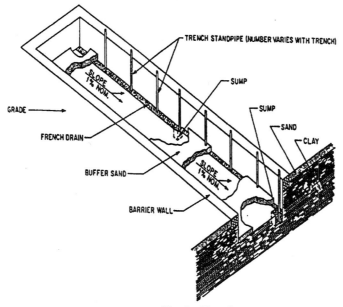

Illustration 1:
Typical Low-Level Radioactive Waste
Burial Trench-Barnwell

To prevent the infiltration of surface water, highly impermeable trench caps cover the completed trenches. With standard trenches, the cap consists of a minimum of at least two feet (0.6 m) of compacted clay over three feet (0.9 m) of soil. In actual practice the overburden ranges from 5-8 feet (1.5-2.4 m). Trench caps are graded to form a gentle slope over the entire burial area. The slope is directed to a central drainageway that carries runoff water to two holding ponds on the western section of the site. Water in the holding ponds either evaporates or slowly seeps into the ground.

LONG TERM CARE

To provide for the long-term maintenance and environmental monitoring of the site after closure, a perpetuity fund has been established with the State of South Carolina. A charge on every cubic foot of waste buried is deposited in a special account. Over the operational life of the site, the principal and interest on these funds will insure that adequate financial resources exist to pay for the long-term maintenance, surveillance, and remediation of the site.

TRIPLE-SAFE DESIGNS

The design of new LLRW sites is significantly different than that used at Barnwell. By incorporating not only federal regulatory requirements but also state, local, and community concerns, Chem-Nuclear has developed the "Triple-Safe" design for waste containment. The design of the Triple Safe system is discussed below.

ENGINEERED BARRIER TECHNOLOGY

Table 3 shows the specific barriers to waste isolation which are employed in advanced LLRW disposal. The first two barriers are part of current practice and will also be employed at future sites. Specifically:

Table 3
LLRW Disposal Site-Waste Isolation Barriers

1. Site Characteristic
2. Waste Form
3. Engineered Barriers
   - Earth Cap
   - Concrete Module
   - Concrete Overpack

- **Site Characteristics** - The disposal site geology and hydrology, in its natural form, must prevent environmental contact with the waste.

- **Waste Form** - The waste generators must prepare the waste in a solid form suitable for secure disposal. The engineered barriers are shown in Illustration 2, 3, and 4.

These functions are as follows:

- **Concrete Overpacks** (Illustration 2) - The waste containers are placed within thick-walled, reinforced concrete containers. A fluid concrete grout is used to encapsulate the waste container within the overpack. Each overpack contains 100-400 cubic feet of waste. Two types of overpacks are used which are cylindrical and rectangular in shape. The intent is to maximize the waste packaging efficiency within the modules.

- **Concrete Modules** (Illus. 3) - Individual concrete overpacks are closely packed within thick-walled, reinforced concrete modules. Each module is closed after being fully loaded with overpacks, and typically contains 50 000 - 100 000 cubic feet of waste. The modules are isolated from ambient conditions

4

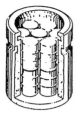
CONCRETE CYLINDRICAL OVERPACKS

CONCRETE RECTANGULAR OVERPACKS

OVERPACKS

All waste will be placed inside concrete overpacks and sealed with cement grout.

Illustration 2:
Concrete Overpacks

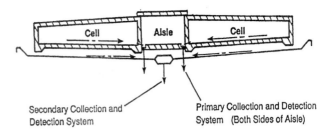

Secondary Collection and Detection System

Primary Collection and Detection System (Both Sides of Aisle)

INFILTRATION COLLECTION AND DETECTION SYSTEM
Located at each end of the center aisle to provide a positive collection and detection system.

Illustration 3:
Concrete Modules

during the loading operations and sealed immediately after loading.

- **Earth Cap** (Illus. 4) - Typically 40-50 modules (in two rows) are grouped and covered with a multi-layered engineered earth cap. The earth cap soil layers passively divert surface water from the concrete modules. The earth cap is composed primarily of natural materials (clay, sand, vegetative soil) obtained from the disposal site or nearby. The modules are completely covered with the earth cap with a cap height extending above the module roof for 7 to 10 feet.

The primary function of the earth cap and concrete modules the diversion of surface and ground water from the waste. The concrete overpacks serve as a barrier in the unlikely event that any water penetrates the waste module. The concrete overpacks, due to their smaller size, greatly facilitate individual inspection

and retrievability (if that is ever needed).

The requirements for the engineered barriers in providing long term, passive, waste isolation is shown in Table II. Long-term experience with the materials comprising the individual barriers does not incontestably prove that any single barrier can provide the requisite isolation function for 300-500 years. For example, concrete as an engineered construction material has a history of less than 200 years. However, as the logic chain shows in Table 4, waste isolation for the required period can be readily proven when the system is taken as a whole--just considering the engineered barriers. Current estimates, using actual waste disposal records, show that the total curie content of the disposed waste is only about 0.1-0.2% of the initial inventory after 300 years (1). Longer-lived isotopes (generally found in Types B, C waste), which comprise most of the remaining inventory, are placed in special, thicker walled - polyethylene lined, concrete overpacks.

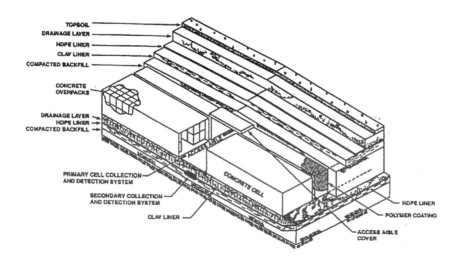

Illustration 4:
Earth Cap

Table 4
Logic Chain

	OPERATION 50 yrs. ACTIVE MAINTENANCE/CLOSURE 10 yrs. ACTIVE INSTITUTIONAL CONTROL 100 yrs.	PASSIVE INSTITUTIONAL CONTROL	200 yrs. INTRUDER CONTROL PERIOD	WATER INFILTRATION ONLY ABOVE GRADE DESIGN 200 yrs.
COVER SYSTEM	LIMITED & CONTROLLED WATER INFLT. (MONITORED)	WATER INFILTRATION ACTIVE STATE MAIN (MONITORED)	NO CREDIT FOR WATER CONTROL	NO CREDIT FOR WATER CONTROL
VAULT	VAULT REMAINS DRY (MONITORED)	NO SIGNIFICANT WATER PENETRATION ACTIVE STATE MAINT. (MONITORED)	POTENTIAL FOR SIGNIFICANT WATER PENETRATION	CREDIT AS INTRUSION BARRIER ONLY
(OVERPACK) RECTANGULAR CLASS A WASTE	NO WATER PENETRATION	NO WATER PENETRATION	POTENTIAL FOR WATER PENETRATION/LEACHING	CREDIT AS INTRUSION BARRIER ONLY POTENTIAL FOR VAULT RELEASE
(OVERPACK) W/LINER CYLINDRICAL CLASS B & C	NO WATER PENETRATION	NO WATER PENETRATION	NEGLIGIBLE POTENTIAL FOR WATER PENETRATION/LEACHING	CREDIT AS INTRUSION BARRIER ONLY POTENTIAL FOR VAULT RELEASE
CONCLUSION	NO VAULT RELEASE	NO VAULT RELEASE	POTENTIAL FOR LIMITED VAULT RELEASE CLASS A ONLY RADIONUCLIDE TRANSPORT ANALYSIS REQUIRED	POTENTIAL FOR LIMITED VAULT RELEASE CLASS A, B, C, & MIXED ONLY RADIONUCLIDE TRANSPORT ANALYSIS REQUIRED

OVERVIEW OF ADVANCED LLRW DISPOSAL SITE

A conceptual design of typical advanced disposal site is shown in Illustration 5. The site can be divided into three general areas:

- **Disposal Areas** - This restricted area contains the waste disposal modules and the waste handling building.
- **Administrative Area** - This area contains the administrative building, laboratories, maintenance buildings, and the waste receipt building, visitor center, site utilities, etc.
- **Buffer Zone/Borrow Area** - This area segregates the disposal area from the site boundary. It also is used for storage of borrow materials used for the formation of the earth cap. Retention ponds are situated for temporary storage of surface water diverted from the disposal area.

These generic regions are common to all such sites and do not specifically identify advanced features. Table 2 presents a comparison of the evolution of the design of LLRW disposal sites over the last 25-30 years and specifically highlights the technology advancements. Of key interest is the current requirements for engineered barriers which makes possible above-ground disposal (well above the natural water table), and the smaller annual quantities of disposal waste. The result of increased confidence in long-term waste isolation and a development of a number of disposal compact sites is increased cost. Typical disposal costs in the 90s (constant dollars) for the new sites are projected at $100-500 per cubic foot. The actual cost will correlate closely with the annual volume of disposal waste. Sites with small volumes of disposal waste will have higher unit costs due to the fixed cost associated with the large capital investment required to develop a new site.

MATERIALS OF CONSTRUCTION

The primary construction material is reinforced concrete. There is considerable experience for use of this material. It provides considerable strength and resistance to environmental factors at reasonable cost. It can also be constructed using recognized standards for quality control and testing, which permits the design to be constructed in a consistent manner and in full accord with the intent of the approved design.
Key elements of the reinforced concrete material must be specified:

- **Concrete Formulation** - Current research on concrete show the importance of low water cement mixtures in ensuring both strength and longevity. Pozzolanic cementitious materials such as flyash are blended with portland cement resulting in both increased strength and increased resistance to chemicals found in surrounding soils; they also provide resistance to the erosion forces of flowing water and freeze-thaw cycles.

- **Reinforcement** - The steel reinforcement bar will not experience harmful corrosion when fully encapsulated in an alkaline cement mixture. As a precautionary measure, the reinforcement will be epoxy-coated to provide increased corrosion protection.

© IMechE 1991 C431/072

Illustration 5:
Conceptual Design-Advanced Disposal Site

The design also makes judicious use of artificial materials (primarily plastic films and coatings) where they will have maximum effectiveness against the effects of possible deleterious chemicals and also provide erosion protection. These materials, primarily polyethylene, are also commonly used in hazardous waste disposal sites. Plastics typically contain additives for UV protection and are buried or otherwise shielded from sunlight. Plastic materials are employed for the following applications.

- **High Density Polyethylene** (HDPE) Liner - The drainage and clay layers of the earth cap are separated by a 2-3 mm (0.080-0.120 inch) liner. This liner is applied in wide sheets and welded at the seams to effect an almost impermeable boundary through the earth cap.

- **Module Coating** - The external concrete surfaces and inner floor of the module is coated with a plastic coating (typically polyurethane) to provide further protection against water erosion.

- **Overpack HDPE Liner** - The concrete overpacks used for class B, C waste incorporate a 10-12 mm internal liner between the poured concrete and grout layers (refer to Illus. 1).
The natural materials selected for construction have engineering properties important to their function. These materials are mostly used in the earth cap (see Illus. 3)

which covers the disposal modules and the foundation upon which the module rests. Properties of particular importance are listed below.

- **Permeability** - The vegetative layer and draining layer must have relatively high permeability (10^0 - 10^2 cm/sec range) to permit lateral movement of surface water. The clay layer must have a low permeability (10^{-7} cm/sec or less) to minimize vertical movement of surface water.

- **Chemical Constituents** - The soil directly abutting the concrete modules will have low chloride and sulfate ion concentrations to minimize chemical interaction with concrete.

As with all nuclear projects, LLRW disposal components will be constructed in accordance with a qualified quality assurance program. The preceding discussion highlights the importance of proper specification of materials and quality control during construction in ensuring that design objectives are met.

SPECIAL DESIGN FEATURES

There are a number of other design features which enhance the operation of new LLRW disposal sites. The design and function of these features are as follows:

- **Waste Handling Systems** - The waste containers are moved from the shipping vehicle, inspected, placed within concrete overpacks, and positioned within the concrete modules. All operations are performed in a closed environment under a negative pressure (as noted in Table 2). As a result, there is no direct means of surface water entering the waste modules during and following loading operations or of any possible radioactive contamination exiting from the handling spaces. The module air is filtered, and monitored. Some sites employ a waste processing building (see Illustration 5), while others use a movable building which directly covers the waste module. The loaded concrete overpacks
  are handled with forklift trucks and overhead gantry cranes. The waste containers are encapsulated by grouting the overpack. This process occurs within the handling building.

- **Test Cells** - Each site will have a representative disposal module designated as a test cell. The cell will have special test instrumentation and inspection areas to monitor concrete performance, waste stability and other technical parameters of interest. The test cell will be loaded during the initial operational period. It will be available for frequent inspection during the entire operational and closure period. Monitoring of cell performance during this period will provide a base line of actual performance. Special areas of concern can be tested to ensure that comparable considerations can be implemented in future operations. In the event that a system design correction is needed, the test cell will serve as precursor of future system performance.

- **Waste Monitoring Systems** - As is current practice, the entire disposal site will have site radiological monitoring systems. These will monitor for radioactivity in the air and groundwater at selected locations throughout the site. In addition, monitoring the actual performance of the waste modules will occur on a continuous basis during the operational closure and active maintenance period. Each module will have a gravity drain system (see Illus. 3) which will collect possible liquid infiltration to the module in a local collection tank. The tanks will be interconnected with piping and directed to a larger monitoring/collection tank. The monitoring tank will be utilized for an entire row of modules. Full-manned inspection access will be provided to both the local and the larger tanks. All collected fluid will be tested for radioactivity. This system provides a direct means of testing the system of earth cap, waste modules, and overpacks for integrity. Fluid flow is by gravity head and the system is totally passive.

- **Site Drainage Systems** - These systems serve to divert rainwater from the waste modules. Subsequently, the system collects the diverted water which is then tested and eventually released off-site. The system consists of rain gutters on the module's roofs with concrete culverts. These culverts direct the water to multi-acre retention ponds located in the site buffer zones. The retention pond(s) are lined and have means of sampling for radioactivity and other chemicals. Tested liquids emanating from the waste monitoring systems would also be released to the retention ponds following testing.

- **Waste Tracking System** - All existing LLRW disposal sites have required and maintained extensive records for waste receipt and disposal. The new LLRW sites will expand these record systems further to ensure that all required data on the waste is tracked from the time it leaves the generators site to its ultimate emplacement in the waste module. This computerized system will permit real time assessment of the on-site isotopic inventory, including the contents within each waste module. Extensive use will be made of bar code technology to track the waste containers as they are handled and move don-site. The proposed systems will not only be beneficial for meeting regulatory requirements but will also assist in planning site operational and construction schedules.

CONCLUSION

New LLRW site design features incorporate multiple barriers to comply with current regulation, adapt to local geologic features and gain public acceptance.

By utilizing modular designs, site development can be tailored to small or large volume waste generation needs.

C431/055

Safety analysis recommendations for the restoration of the Richard near surface repository in Czechoslovakia

M VANECEK, MSc
Centre of Mining Hydrogeology, Geoindustria GMS, Prague, Czechoslovakia
L NACHMILNER, PhD
Nuclear Research Institute, Rez, Czechoslovakia

SYNOPSIS On the territory of Czechoslovakia three deposit sites for institutional radioactive wastes without corresponding geological investigation were selected in the past. Activity of all these disposal structures respected the criteria acceptable at its period. At the present, the operator efforts to take these facilities to accordance with actual I.A.E.A. recommendations. Still in-run disposal system situated to old drift mine Richard II close to city Litoměřice and discussed here is one of them.

1. INTRODUCTION

In mid-1989, the operator of the Richard II Radioactive Waste Deposit Institute for the Research, Production and Applications of Radioisotopes felt that it was necessary to assess the safety of biosphere in the event of a leakage of radioactive pollutants into groundwater system with which the mine complex is communicating. It was agreed that the input would be represented by archival data processes using a computer, and contain no new technical or analytical works. At the same time, the client required the resulting report to contain a tentative outline of further steps possibly needed to safeguard groundwater against pollutants as well as an indicative assessment of the abondoned mine stability.

2. HYDROGEOLOGICAL ENVIRONMENT

The area of interest is a part of the North Bohemian Cretaceous Platform, the sequences of which are subhorizontal and dipping slightly south-south-westward here. The Upper and Middle Cretaceous sediments is represented by developed Cenomanian and Lower Turonian layers of sandstone and clayey sandstone. Overlying these are continuous Middle Turonian sediments consisting of fine-grained sandstone passing into claystones. Upper Turonian strata consists from marlite intercalated limestone. Quaternary slightly sandy loams are covering most of the area and slopes of significant Tertiary volcanic hills are covered with loamy debris.

Lithological and facial diversity of the site permits a vertical division of the sedimentary sequences into hydrogeological aquifers and aquicludes which differ in several orders of magnitude with respect to their hydraulic parameters. There are generally three aquifers separated with aquicludes off pelitic sediments. The main one is associated with Cenomanian and Lower Turonian rocks and denoted as the basal Cretaceous aquifer. The middle one occurs in some places only and is usually developed at the upper part of Middle Turonian porous sediments. The upper aquifer occurs in the relatively thick Quaternary deposits here.

Tectonic faults have divided the area of interest into rock blocks with a certain amount of uplifting. Some geological reports from he region mention vertical throws as high as 70 m. As far as the underground structure of the Richard Mine property is concerned, Turonian sediments are pervaded almost exclusively by a system of subvertical to vertical faults. Most of the faults and fissures in the statically undisturbed rock mass are closed. Occasional open fissures have been completely healed with limestone or gypsum.

In the past there have been several geological surveys carried out in the broader surroundings of the Richard Mine which were aimed at regional prospecting for deposit indications, investigation of a deposit of calcareous marlstones and groundwater resources. They have provided a sound idea of geological and hydrogeological conditions on a regional scale. However there has been no geological or hydrogeological survey carried out on the site. This statement is contradictory to the presented geological cross-section.

3. DESCRIPTION OF THE MINE COMPLEX

Marlstone with a higher content of calcareous component have been used as an input material for the nearest Cement Works (off Lovosice, North Bohemia) since 1895. The exploitation ended in 1962 following a negative geological report.

The most extensive and southernmost mining field bears the name of Richard I. The underground complex axis is a NW–SE 639 m long main gallery with a surface mouth in the southeastern part. One part of Richard I mine were converted into an underground factory during the Nazi occupation. The postwar period witnessed renewed mining activities which have resulted in doubling the underground space.

The part of the underground complex situated northeast of Richard I is called Richard II. The two parts are connected by a NE-SW, 150 m long gallery. Richard II has its own entry and was used as an underground factory under the Nazis as well. Most of its underground space was rebuilt during the sixties and converted into a deposit of radioactive wastes, except those produced by power generation facilities. The working is sunk approx. 420 m into the hill. The main access gallery runs WNW and there is a ventilation shaft at its end. It has partly been walled over: the shaft is completely filled with debris and its sides are collapsing.

The name Richard III denotes the northernmost, triangular underground space one apex of which is connected to the ENE part of Richard II. One of the two remaining apexes is a caved-in entry. In all available archival documents, Richard III is classified as inaccessible and this is about all information there is.

The disposal area falls into the former underground factory of the Richard II mine. For the purpose of the radioaktive waste deposit, the gotten underwent some modifications during the sixties. Ceilings of adits and galleries were in some places provided with safety steel meshing anchored to the rock. A new water-collecting drainage was built under the floor of the main access gallery, which drains water infiltrating into the main close to the deposit. Spaces filled with radioactive wastes have been walled over. A small quantity of unidentified wastes dating back to the Nazi occupation may also have been deposited here.

In the rearmost part of the Richard II gotten, there is an inaccessible pillar of a vertical mine working which has allegedly been used as a ventilation shaft. Drainage holes run below the floor level, the purpose of which is to take groundwater from the debris-slogged shaft away. The groundwater is then carried into catchment troughs. The front trough accumulates water and carries it under the main gallery floor away from the mine complex, emptying into the sewage system of the city of Litoměřice close to the access road.

We believe that the genesis of current mine water differs from the historical one. In all probability, historical infiltration was also attributable to the vertical mine working passing through a water-bearing tectonic fault. An analysis of facts available suggests that the Richard II mine was historically a much greater producer of mine water than it is now. The current situation may result from the inaccessible ventilation shaft having acted as a drainpipe for a protracted period of time, as well as from the relative lowering of the discharge head of the basal Cretaceous aquifer in the Píšt'any meander of the Elbe, or a combination of both. The fact is more than crucial for any attempt at predicting the hydrogeological evolution of the site and the region, and would require an ad hoc hydrogeological survey.

4. DISPOSED RADIOACTIVE WASTES

Radioactive wastes have been disposing in the Richard II facility for twenty years. During that time, their forms and radionuclide composition have varied. According to available data, the waste deposit facility can be characterized as follows:

1. Waste forms – most of the radioactive wastes have been cast in bitumen, a smaller part in cement. The product is either homogeneous, or consists of contaminated solids cast in an active or inactive cement mixture. Materials which have not been fixed in the above way are usually uncompacted, rarely packed in PE bags. Some enclosed sources of radiation have not been fixed.

2. Containers – 60-litre containers, later 100- and 200-litre galvanized steel barrels have been used. Some of them have been corroded through, but most of them are in quite a good shape, especially those with a protective coating. Enclosed sources of radiation have been partly stored in lead containers.

3. Microclimate – the temperature in the mine is stabilized at 10°C throughout the year, the relative humidity in open spaces is also constant – 90%. No humidity measurements have been performed in closed spaces, but the humidity there may be expected to be close to that of saturated vapours.

The simplified hydrogeological situation permitted us to determine the relative criticalities of the different radionuclides in terms of the safety of people using contaminated water as their staple drinking water.

The calculation has shown that there will be no unacceptable personal exposure, as the dose increment is lower (in the order of magnitude) than fluctuations of the natural radioactive background in Czechoslovakia (100-200 Sv.a^{-1}). In addition, the results presented in the table clearly indicate that ^{14}C and ^{3}T isotopes pose the highest risk in terms of groundwater contamination. Owing to the decay period of the ^{14}C isotope, the risk persists for several tens of thousands of years.

5. RESULTS

Based on facts submitted in 1st. stage of Safety Analysis Report the following conclusions can be made:
- the site affected by the underground working of the Richard Mine has not been explored sufficiently in terms of its geological and hydrogeological mode,
- there is no monitoring system on the site,
- only the used part of the underground space has been rehabilitated,
- if there is a disturbance of the geotechnical stability of non-rehabilitated parts of the mine, kirved joints may reach up to the surface and result in a change of hydrogeological conditions,
- there is a free access from the surface into the mine and with a bit of effort to the waste disposal area as well,

cells filled with radioactive wastes are closed by a brickwork partition which increased humidity condensing as a continuous layer of water on the floor, in order to reduce the number of air shafts to be checked the circulation of air in the mine has been curtailed which has increased the humidity of the underground working and swelling of the pelitic sediments,

The existing information prompts the following conclusion: the existing radioactive waste repository may be operated conditionally until the time the facts outlined above are confirmed by an ad hoc hydrogeological survey.

It should be mentioned that we were able to observe changes in the stability of the underground working since 1989, which were manifested by an increased amount of rock fragments fallen on the floor and by increasingly difficult accessibility of remote parts of the mine. As kirved joint drags reach up to the surface the underground structure is subjected to increased groundwater inflows which may even peak in floods. Any disturbance of stability in any part of the Richard Mine complex, which is sufficiently sealed, is intolerable. The duration of the irreversible process may be estimated at 100 years or so but no longer, a time incomparable to that of toxicity of materials deposited there. The kirved joints will provide a connection between the surface and the deposit, draining surface and near-surface runoffs. In the course of time, free passages of mine water flow will be blocked, the water will accumulate in the underground space and leach the deposited radioactive wastes. The duration of the process can be estimated at several hundred years.

The used space of the abandoned mine is presently regarded as a shallow ground repository of radioactive wastes produced by non-energy operations. All deposits must comply with criteria specifying the stability of disposed materials, properties of disposable areas, functions of natural and technical barriers. A functional monitoring system is checking engineering conclusions in reality and is necessary.

6. PROPOSED COURSE OF ACTION

We have prepared several alternative solutions and measures aimed at protecting the biosphere of the broader surroundings against adverse effects of radioisotopes deposited in the mine namely:

1. To abandon the deposit site completely and to deposit any wastes hitherto disposing there in a repository facility for radioactive wastes produced by nuclear plants.
2. To undertake geological exploratory works aimed at confirming conclusions of the present report. These should result in a tentative safety analysis. If the analysis is negative, measures outlined under Paragraph 1 will have to be taken. If positive, the procedure should be as follows:
3. To rehabilitate the mine to a distance where a "sealing area" separating the radioactive waste deposit site from unused underground spaces could be built. The distance of the sealing area depends on the range of kirved joints.
4. To build a "sealing wall" separating the used part of the mine space from the unused one. The sealing wall should be impermeable (waterproof) in both directions.
5. As soon as the underground space and its different compartments have been filled with wastes, the entries should be sealed by means of pressure-resistant curtains and the space around the wastes should be grouted or packed with a natural water-impermeable non-decomposing material. The above procedure will have to be used even when the existing waste deposit facility is not expended.

REFERENCES

(1) LASTOVKA, J., NACHMILNER, L., VANECEK, M. A Richard II - safety analysis - 1st stage, 1990, 38, RE.S.A.

A Review of Radioactivity of Materials Deposited Since 1979 Till 1988 in Richard Disposal Site

Nuclide	T 1/2 (yrs)	Act. 1 (Bq)	Act. 2 (Bq)	Act. 3 (Bq)	Rd
^{106}Ru		5.6 E10	3.2 E07		
^{109}Cd		2.1 E10	1.1 E09		
^{110}Ag		4.4 E08			
^{113}Sn		9.2 E09			
^{125}J		5.4 E12	7.0 E08		
^{133}Ba		2.0 E10	1.0 E10		
^{134}Cs		4.2 E09	1.0 E09		
^{137}Cs	3.0 E01	4.3 E12	3.7 E12	7.4 E10	110
^{144}Ce		9.4 E10	1.7 E10		
^{147}Pm		5.5 E12	1.2 E12		
^{14}C	5.6 E03	9.0 E12	9.1 E12	1.8 E09	
^{192}Ir		2.5 E11	9.9 E08		
^{203}Hg		3.9 E11	1.0 E08		
^{204}Tl		1.2 E12	3.0 E11		
^{210}Pb		8.7 E09	7.6 E09		
^{210}Po		5.9 E10	2.9 E09		
^{226}Ra	1.6 E03	5.8 E10	5.8 E10	1.2 E05	300
^{22}Na		1.4 E11	1.7 E10		
^{238}Pu	2.4 E04	8.9 E11	7.5 E11	1.5 E06	600
^{238}Pu		3.3 E09	3.2 E09		
^{241}Am	4.5 E02	9.1 E13	9.1 E13	1.8 E07	300
^{36}Cl		9.2 E10	9.2 E10		
^{3}H	1.2 E01	9.3 E12	6.2 E12	1.3 E10	1
^{45}Ca		1.8 E11	4.9 E09		
^{54}Mn		4.0 E10	8.2 E09		
^{55}Fe		1.3 E10	4.5 E09		
^{57}Co		1.2 E11	3.1 E10		
^{59}Fe		1.1 E12			
^{60}Co	5.3	9.2 E13	8.9 E13	1.8 E08	200
^{63}Ni		1.8 E09	1.7 E09		
^{65}Zn		1.5 E11	4.1 E09		
^{75}Se		2.4 E11	4.6 E10		
^{85}Kr		4.9 E10	4.0 E10		
^{88}Y		9.1 E10			
^{90}Sr	2.8 E01	9.2 E13	9.1 E13	1.8 E11	330
		3.1 E14	2.9 E14		

Explanatory notes:
Act. 1 - initial radioactivity
Act. 2 - radioactivity calculated as 18th of December 1989
Act. 3 - leached radioactivity

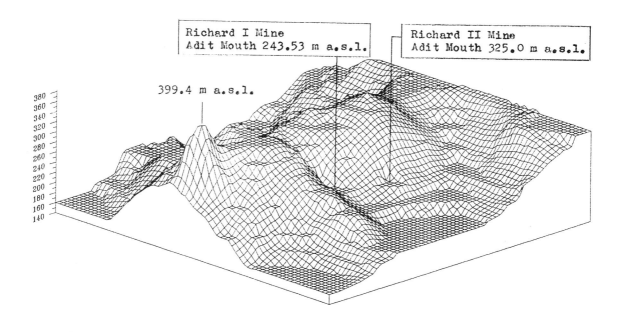

Picture No. 1: Surface of Richard Gotten Site

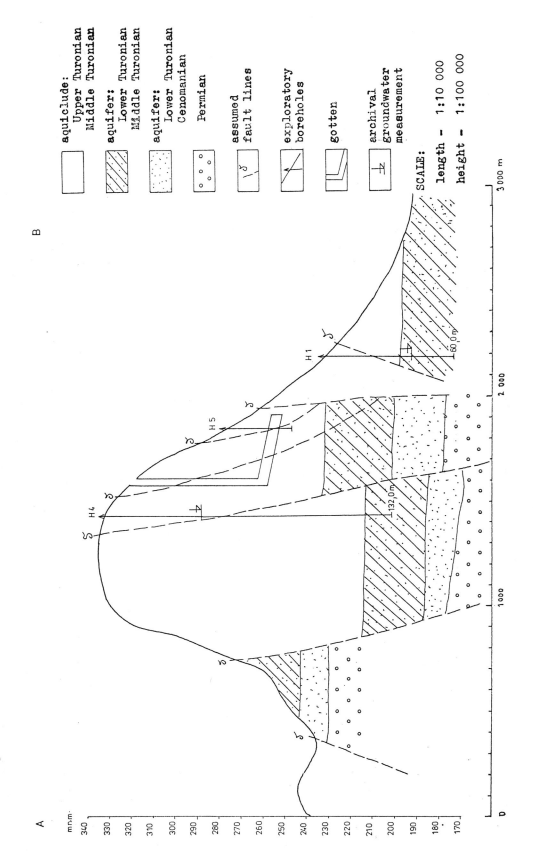

A Superelevated Crossection of the Hydrogeological Structure

aquiclude:
Upper Turonian
Middle Turonian

aquifer:
Lower Turonian
Middle Turonian

aquifer:
Lower Turonian
Cenomanian

Permian

assumed
fault lines

exploratory
boreholes

gotten

archival
groundwater
measurement

SCALE:
length - 1:10 000
height - 1:100 000

C431/026

The management of radioactive wastes at BNFL, Sellafield

W HEAFIELD, BSc, CEng, FIChemE, S G DONN, BSc, CEng, MIChemE and
H G LEWIS, BSc, PhD
British Nuclear Fuels plc, Risley, Warrington, Cheshire

SYNOPSIS   BNFL's policy for managing radioactive waste at its Sellafield site is to minimise effluent discharges to the environment, to dispose safely solid low level waste as it arises and to provide safe and cost effective methods for storing, treating and preparing for disposal all other wastes. In pursuance of this policy BNFL has progressively and significantly reduced activity levels in liquid effluent discharges and is building a new actinide removal plant. BNFL has also designed, built and is currently operating intermediate and high level waste treatment plants; further plants for intermediate and low level wastes are under construction. The strategy employed for waste management is described in this paper together with its development against a background of statutory legislation and increasing public awareness.

INTRODUCTION

Fuel cycle operations including reprocessing have been carried out at the British Nuclear Fuels plc (BNFL) Sellafield site since 1952. The reprocessing operations produce a range of radioactive wastes and over the past 40 years BNFL has developed an increased understanding of waste composition and conditioning requirements against a background of statutory legislation and increasing public awareness.

BNFL has adopted a policy for management of radioactive waste whereby effluent discharges to the environment are minimised, solid low level waste is disposed of as it arises, and all other wastes are stored, conditioned and treated for eventual disposal. This paper discusses the strategy that has been used to implement this policy, reviews progress towards the implementation of the strategy and explains some of the factors which have influenced that strategy.

WASTE CATEGORIES

In the United Kingdom it is common practice to allocate radioactive wastes into one of three categories.

High Level Wastes (HLW), or heat generating wastes, are defined as those in which the temperature may rise significantly as a result of their radioactivity. The design of storage or disposal facilities for HLW must therefore take this factor into account.

Low Level Wastes (LLW) contain radioactive materials not acceptable for disposal as ordinary non-radioactive trash, but not exceeding 4 GBq/te alpha or 12 GBq/te beta/gamma.

Intermediate Level Wastes (ILW) are those containing radioactivity above the levels set for low level waste, but which do not experience a significant temperature rise as a result of their radioactivity.

In practice the HLW category is applied exclusively to the concentrated waste product from the primary separation stage of fuel reprocessing. Consequently, in adopting this classification certain wastes are categorised as ILW, although they may generate a modest amount of heat.

NATIONAL STRATEGY AND LEGISLATIVE ARRANGEMENTS

It is Government policy that wastes should be disposed of under strict supervision to high standards of safety and the periods of storage should be the minimum compatible with safe disposal. For heat-generating wastes this means storing the wastes for at least 50 years to allow radioactive decay to reduce the heat rating substantially. For low and intermediate level wastes it is desirable to dispose of them as soon as possible to avoid the creation of additional accumulations and the provision of costly and extensive storage capacity.

The principal objectives include:

-    ensuring that the creation of waste is minimised;

-    ensuring that the handling and treatment of wastes are carried out with due regard to environmental considerations;

-    securing the programmed disposal of waste (already) accumulated at waste producers' sites.

The basic radiological protection principles on which the strategy is based are derived from the work of the International Commission on Radiological Protection and are embodied in the various legislative instruments. They include the following:

-    all practices giving rise to radioactive wastes must be justified, ie. the need for the practice must be established in terms of its overall benefit;

-    radiation exposure of individuals and the collective dose to the population arising from radioactive wastes shall be reduced to levels which are as low as reasonably achievable (ALARA), economic and social factors being taken into account;

-    the average effective dose equivalent from all sources, excluding natural background radiation and medical procedures, to representative of members of a critical group should not exceed 1 mSv per year.

As part of its own standards BNFL has adopted a design limit of 15 mSv per year as the maximum individual operator dose for new plant, with a design target averaging 5 mSv for the plant workforce as a whole. Both these levels are subject to the application of ALARP (as low as reasonably practicable) not requiring doses to be reduced further.

HIGH LEVEL WASTE MANAGEMENT

Highly active liquor (HAL) arising from the first separation cycle in spent fuel reprocessing normally contains 97-99% of the fission product activity. This liquor is concentrated by evaporation and is stored in high integrity, shielded and cooled tanks. There are at present some 1400m³ of HAL stored at the Sellafield site. Future reprocessing operations will result in a significant increase in HAL levels, and it is estimated that by about 2005 (after approximately 12 years operation of the THORP reprocessing plant) HAL arisings will have totalled about 3300m³.

While liquid storage has been demonstrated to be safe and reliable with more than thirty years operational experience, storage in a solid form significantly reduces the potential for release of radioactivity and the waste in this form can be suitable for eventual disposal.

BNFL has therefore decided to immobilise the HAL by incorporation into borosilicate glass, thereby improving safety and handling convenience and providing a product suitable for storage, transport and ultimate disposal. The vitrification of HAL at Sellafield in the Windscale Vitrification Plant (WVP) commenced in mid-1990 and in accordance with Government policy the canisters of glass produced are being stored in a natural ventilation air-cooled store where they will remain for at least 50 years.

The vitrification process consists of four basic steps:

1. Metering the radioactive liquors into a heated rotary calciner where water and acid are evaporated to leave a dry powder in which the waste product nitrates are converted to oxides.

2. Mixing these oxides with measured quantities of glass frit in an induction-heated melter which converts the mixture into molten glass.

3. Casting the glass into 150 litre (about 400 Kg) stainless steel containers.

4. Welding on the container lid, decontaminating and placing the container in the adjacent store.

The plant has been designed to allow remote replacement of both calciner and melter units and remote maintenance of container handling equipment. Demonstration of, and practising for these operations played an important part in the preparation and commissioning of the plant prior to active operation.

WVP, which has two process lines, has sufficient capacity over its operating life to treat the arisings from both the Magnox and THORP reprocessing plants as well as the backlog of stored HAL.

A recent detailed assessment by the independent UK Advisory Committee on Safety of Nuclear Installations (ACSNI) has validated the design adopted by BNFL.

INTERMEDIATE LEVEL WASTE MANAGEMENT

Although a large number of ILW streams are generated at BNFL's Sellafield site, they may be grouped according to their origins or properties as follows:

- Fuel Element Cladding Wastes: comprising essentially solid wastes such as leached cladding and end assembly fittings from oxide fuel (known as hulls) and Magnox fuel element cladding (swarf).

- Miscellaneous Beta-Gamma Wastes: waste which, although contaminated, normally has relatively little alpha emitting activity associated with it. This might typically be maintenance scrap and certain spent fuel storage equipment.

- Slurries: a broad range including ion exchange materials, flocs from liquid effluent treatment and sludges from the corrosion of Magnox swarf in water.

- Plutonium Contaminated Material (PCM): alpha-contaminated waste, eg. drummed packages of small items and large pieces of equipment such as redundant gloveboxes, arising from plutonium processing plants. Beta/gamma activity is normally relatively low.

Historically, ILW has been stored at BNFL's Sellafield site in its raw, unconditioned form. Approximately 30,000m³ of ILW have arisen from nuclear fuel handling and reprocessing operations at the site over the past 35 years. As many of the storage facilities currently available for ILW do not have the capacity to accommodate future arisings, BNFL has been faced with the choice of building further storage facilities for unconditioned wastes or conditioning (immobilising) the waste in anticipation of eventual disposal. While raw waste storage generally meets current safety regulations, detailed technical assessments have shown that storage in an encapsulated form would be preferable.

BNFL has therefore developed the strategy that future ILW arisings will be treated, as they arise, by direct encapsulation and that no further raw waste storage facilities will be constructed. In addition, existing wastes will be retrieved and conditioned on a timescale determined by the safety and condition of their respective stores, encapsulation plant capacity and repository availability. This strategy not only improves overall safety on the site but is also less costly than the alternative of providing additional raw waste storage followed by retrieval and conditioning at a later date, as it:

- reduces the mobility of activity, thereby reducing the consequences of a major event such as fire or aircraft impact;

- reduces the requirements for monitoring and surveillance of waste;

- reduces operator dose uptake by reducing the number of handling operations;

- minimises store contamination and thus reduces decommissioning costs.

IMPLEMENTATION OF ILW STRATEGY

There are, at present, no statutory acceptance criteria for radioactive waste disposal packages in the United Kingdom, and given that waste disposal is dependent on the multi-barrier concept, it is believed that over-prescriptive requirements for any one of the barriers, and this applies particularly to the waste package, may not be appropriate. Nirex is currently developing package specifications which will take account of relevant information emerging from its detailed site investigation and repository design programmes. Against this background and in view of the long lead times for designing and constructing encapsulation plants, BNFL has found it necessary to formulate its own package specifications. Waste encapsulation processes and equipment have been developed and designed to meet these specifications.

Since 1982 BNFL has carried out a detailed programme of development work (part-funded by the Department of Environment) aimed at selecting encapsulation matrices for ILW streams, and investigating and defining the properties of the selected encapsulated waste product. This programme has extended over four phases :-

1. Characterisation of the waste, identification of possible encapsulants and selection of several preferred encapsulants.

2. Experimental investigation and comparison of the potential encapsulants; selection of a single preferred encapsulant using multi-attribute decision analysis and taking account of all the waste management stages including disposal.

3. A systematic detailed investigation of a wide range of product properties for the selected encapsulant.

4. Assessment of long term product properties and investigation of the acceptable operating regime which could be tolerated in an actual encapsulation plant.

This formal, structured approach has resulted in the selection of various cement formulations for encapsulation of all the waste streams studied. These formulations, based on Ordinary Portland Cement (OPC) and either Blast Furnace Slag (BFS) or Pulverised Fly Ash (PFA), have been shown to provide the optimum product. The encapsulation process is relatively simple, flexible and operates at low temperature to produce a durable, strong, monolithic product having excellent retention characteristics for actinides.
Nirex generally has been supportive of BNFL's strategy, and, specifically, has carried out detailed evaluations of those aspects of BNFL documentation which are concerned with waste package characteristics for disposal operations. As a result of these evaluations, Nirex has been able to endorse BNFL's approach thereby providing important support to BNFL's submissions to the Regulatory Bodies for permission to construct and operate the relevant plants at Sellafield.

Cement encapsulation of ILW streams will take place in four plants one of which, for Magnox swarf, commenced operation in 1990; the others are in course of design and construction.

ENCAPSULATION PLANT 1 (EP1)

EP1 has been built to encapsulate Magnox swarf and started active operation in mid 1990. This is a single line, multi-stage plant in which the waste is tipped into a 500 litre stainless steel container, water is removed and an inactive pre-mixed cement grout is added whilst the container is vibrated to remove air and assist flow of the grout into voids in the waste. After the initial grout set the ullage is filled with a capping grout and a lid fitted. This lid incorporates a filter to allow any gas generated to escape. After checking for external contamination and, if necessary, decontamination, the containers are transferred, four in a stillage, to an adjacent store.

The plant, which has a nominal capacity of 2000 drums per year, relies heavily on in-cell handling equipment to move drums through the various processing stations. The reliability of this equipment is, therefore, a key factor in the plant's availability.

ENCAPSULATION PLANT 2 (EP2)

EP2 will treat THORP waste streams, both solid hulls and slurries. A similar in-drum grouting process to that employed in EP1 will be used to encapsulate hulls in 500 litre stainless steel containers.

For the encapsulation of THORP waste slurries, dry cement powder will be added to the waste in a 500 litre stainless steel drum which contains a paddle to mix the waste and which will be encapsulated along with the waste. Construction of EP2, which has a nominal capacity of 4000 drums per year, is well advanced and the plant is scheduled to commence operations at the end of next year.

WASTE PACKAGING AND ENCAPSULATION PLANT (WPEP)

WPEP is being constructed to encapsulate flocs from the treatment of liquid effluent and MA concentrate which will originate from the new actinide removal plant (EARP). As with EP2 slurries, dry cement powder will be added to the waste and mixed with a sacrificial paddle; the plant also has the facility to grout certain miscellaneous solid wastes. This plant, which has two lines (one designated stand-by) and a nominal capacity of 1800 drums per year, is also due to be commissioned in late 1992.

WASTE TREATMENT COMPLEX (WTC)

The design of WTC is being reviewed with the aim of providing an encapsulation facility for PCM. Based on the current concept 200 litre drums containing packaged wastes will be supercompacted before encapsulation in cement in 500 litre stainless steel containers. The plant, as currently envisaged, could be operational in 1996.
In addition to these four encapsulation plants, solid beta-gamma waste is now being sorted, packaged and stored in 3.5m³ cement lined steel boxes in the Miscellaneous Beta-Gamma Waste Store (MBGWS); this also started active operation in 1990.

The operation of the plants described above will allow the total volume of raw ILW to be reduced to about half current levels over the next 15 years by which time conditioned waste stocks will be approaching 100,000 containers (drums and boxes).

Packaged wastes produced by the encapsulation plants will be stored in a range of dedicated storage facilities, each optimised to suit specific process and safety requirements. The first store to commence active operation, EPS1, has a capacity of 12,000 drums. Stillages are stacked 16 high in channels, to ensure cooling and ventilation, by a charge machine arrangement which provides the necessary shielding and loading precision.

LOW LEVEL WASTE MANAGMENT

Sellafield is the major generator of solid LLW in the United Kingdom, and accounts for approximately 85% of estimated future arisings. To date the majority of solid LLW generated in the UK has been disposed of at the Drigg site which is owned by BNFL and has operated since 1959. Originally solid LLW was placed in sealed bags, tipped into open-cut, carefully prepared clay based trenches, which were then covered with a layer of earth. However, several important improvements in the management of this waste have been introduced, including volume reduction, waste containerisation and the implementation of a rigorous quality assurance and checking regime.

The most significant improvement has been in terms of visual impact due to the move to engineered concrete-lined vaults rather than trenches. These vaults, which incorporate surface, perimeter and under-slab drainage systems, have been in use for the past three years. When full, the vaults will be closed using engineered, water shedding caps.

The ongoing upgrade of Drigg operations is supported by an extensive development programme aimed at providing the methodology and data for thorough post-closure safety assessments. The programme includes studies into waste degradation, leaching, materials ageing and chemical modelling, geological and hydrogeological studies, groundwater flow modelling and risk assessment methodology development.

BNFL is now carrying out detailed design of a plant to be built at Sellafield for the supercompaction of solid LLW. This will serve to minimise voidage thereby improving the quality of the Drigg repository and extending its operating life. The centralised compaction and containerisation plant is planned to be operational by 1993/94. The waste, in half height ISO freight containers, will be transported to Drigg for grouting and disposal in vaults. Reduction of LLW arisings from better practices and compaction, coupled to improved vault design, indicate that the currently consented area of Drigg should have sufficient capacity to approximately 2050.

CONCLUSIONS

BNFL has developed waste product specifications which have enabled a waste management strategy based on early encapsulation of wastes to be implemented at its Sellafield site.

The Windscale Vitrification Plant is now operating to immobilise highly active liquid wastes in a borosilicate glass. Another plant, EP1, is encapsulating intermediate level wastes in a cement matrix and further ILW encapsulation plants are being designed and constructed. Ultimate disposal of solid ILW and LLW is planned to be provided in a deep underground repository which is currently being planned and investigated by UK Nirex Ltd.

Improved management practices are being introduced at the Drigg low level waste disposal site, extending the estimated operating life well into the next century and enhancing the long-term safety and visual impact of disposal operations.

C431/069

Low level waste management and dose reduction programmes in PWRs

M F P DUBOURG, MS
Framatome, Paris, France

SYNOPSIS   The major part of the French Nuclear Program consists of standardized 900 MWe and 1300 MWe PWR designed and built by FRAMATOME and operated by Electricité de France (EdF).   An extensive effort has been devoted for many years by EdF and FRAMATOME for reducing the volumes of dry actived wastes and technological wastes resulting from maintenance work and plant operation to be sent to surface sites for permanent storage. In addition an important effort for reducing Occupational Radiation Exposure (ORE) is undertaken.

1   INTRODUCTION

In 1990, the fraction of electricity generation produced by nuclear power plants was about 75%. This generation was produced mainly by PWR's. In France, in 1990, 34 units of 900 MWe class and 17 units of 1300 MWe class were in operation.

The low level radioactive waste volume which was at a level of 375 m^3 per plant and per year in 1985 was reduced to a level of 180 m^3 per plant and per year in 1990.

Operating experience with 900 MWe PWR plants and 1300 MWe PWR plants in France and in Belgium has shown that average collective doses during plant operation and maintenance are lower than those recorded in other foreign countries including United States of America.

This is mainly due to the enhancements in FRAMATOME design engineering and the Electricité de France's operating procedures have minimized personnel exposure during maintenance and repair

On average in 1988 the annual dose for a French plant was about 180 man rem (50 units) and annual doses for US plants were respectively 346 man rem for PWR's (70 units) and 510 man rem for BWR's (36 units).

This good result in annual dose is related to the precise control of the circuit contamination and to the extensive use of remoted tooling for plant maintenance and inspection.

2   TREATMENT AND PROCESSING OF THE EFFLUENTS AND WASTE PRODUCED BY 2 PWR UNITS OF 900 MWe

2.1   Generalities

During normal operation a twin 900 MWe PWR produces liquid and gaseous radioactive effluents which must be treated, accounted and discharged as well as solid radioactive waste which must be processed before long term storage.

The waste treatment system is given in the schematic overview and is common to both units (see fig. 1).

It enables liquid and gaseous radioactive waste to be collected and treated so that it can be released to the environment whilst complying with the ICRP rules for the protection of plant workers and local population.

Prior to treatment, waste is collected and often stored. Treatment varies according to the waste activity level. The vent and drain system (RPE and SRE) collect and route effluents to the appropriate treatment system.

The boron recycle system (TEP) separates the non-contaminated coolant effluents into boric acid solution and reactor grade water, both to be reused as make-up to the reactor coolant system, thus limiting the amount of effluents to be discharged to the environment.

The liquid waste treatment system (TEU) incorporates 3 functions :

(a) demineralization,
(b) evaporation,
(c) filtration.

Purified waste is released from the plant after monitoring. Should release conditions not be met (low dilution capability of the natural environment or too high radioactive level of the waste), they can be stored in the liquid waste discharge system (TER) and recycled for reprocessing.

Radioactive concentrated waste is treated by the Solid Waste Treatment System. The gaseous waste treatment system (TEG) reduces effluents activity level before discharge to the plant stack and release to the atmosphere. It functions either as a filtration system (aerosols, iodines) or as a tank storage system for purpose of radioactive decay.

2.2 Dry waste treatment conditioning

The solid waste treatment system (TES) is common to both units. It is located partially in the BAN (Nuclear Auxiliary Building) partially in the WAB (Waste Auxiliary Building)

It is designed to collect, store, process and package radioactive waste before shipping it to a long term storage facility, LA MANCHE or AUBE Repository.

This waste includes 4 main categories :

(a) ion-exchange resins,
(b) evaporator concentrates,
(c) filters,
(d) miscellaneous waste (tools, vinyl bags, rags, clothes ...).

The solid waste treatment and conditioning system (TES) has been designed and carried out by the PEC Engineering Company. The treatment consists of using a concrete process for the 3 first categories. The miscellaneous waste being simply packaged in metal drums with compacting where appropriate.

A geographical separation between the concrete batching plant and the nuclear auxiliary building has been designed to avoid dust transfer (from cement, sand, lime) to the rooms.

Each waste transfer to the TES is carried out under radioprotection, as follows :

- Hydraulic transfer for resins and pneumatic transfer for concentrates by means of protected pipes,

- Filter transfers by means of a special lead cask.

The main advantages of the process are simplicity, low investment and operation costs and security of radiological protection.

3 WASTE TREATMENT AND PACKAGING (see fig. 2)

The first 3 types of waste are conditioned in the drumming station in the BAN, level 0.

A concrete batching plant, adjacent to the waste auxiliary building, carries out the preparation of concrete batches, by mixing cement and sand in the right proportions.

2 types of batches are prepared :

(a) dry batches for the drumming of resins and concentrates,

(b) wet batches for the drumming of filters.

These batches are transferred to the BAN by means of skips. They are poured at the upper level of the BAN in vertical tubes where they fall into the concrete drums.

Dry loads are then mixed with the waste, inside the drum, by means of a disposable blade.

The filters are embedded in liquid concrete. A vibrating table is used to improve the filling of concrete around the filter.

After they have been filled, drums can receive a prefabricated cap to allow temporary radioactive protection. After a setting time of about one day drums are transferred to the waste auxiliary building for final capping.

This final capping is carried out by means of a belt conveyor transferring concrete from the concrete batching plant to the final capping station. Concrete is poured into the upper part of the drum for radioprotection completion. A vibrating table is used to improve the filling of concrete.

After a minimal storage time of one month, drums can be shipped outside the plant for final storage.

Radioactivity monitoring is carried out before shipping to ensure that the drums comply with the rules of the transportation of radioactive materials.

The fig. 3 gives the decrease of the volumes of Low Level Dry Active Waste produced by a typical PWR in France.

4 DOSE REDUCTION PROGRAM

Enhancements in FRAMATOME design engineering and the Electricité de France's operating procedures have minimized personnel exposure during maintenance and repair.

Fig. 4 shows the annual operational dose versus energy output for US plants and various French PWRs. For instance, each TW.h of energy produced by French PWRs corresponds to an annual dose of 33 man rem. for the same quantity of energy generated in the USA, the annual dose is 62 man rem for PWRs and 117 man rem for BWRs.

FRAMATOME has implemented the following measures to reduce Occupational Radiation Exposure (ORE) as part of its R&D programme :

(a) Reduction of the source and quantity of corrosion products susceptible to activation.

(b) Reduction of maintenance personnel exposure time through adequate training and improved maintenance tooling.

4.1 Reduction of amounts of activable materials

Two types of materials have to be considered :

(a) In flux materials, e.g. for fuel assemblies and core support structures.

(b) Primary system materials, e.g. for steam generator Inconel tubing and hard faced materials used for valves, control rod drive mechanisms and centring keys.

4.1.1 Material submitted to neutron flux

Stainless steel and Inconel, materials exposed to neutron flux within the reactor, contain residual amounts of cobalt.

© IMechE 1991 C431/069

Specifications restrict the cobalt content to 0.080 %, but in most cases, the actual cobalt content in materials exposed to neutron flux is lower than the specified amount.

In fuel assemblies, ^{58}Co is produced by activation of Inconel grids. In the FRAMATOME developed advanced fuel assemblies (AFA), Inconel grids are replaced by Zircaloy grids with Inconel clip springs. The Pactole program computations show potential reductions in ORE of up to 30 %.

Another way to reduce ORE is by improving the high burnup performance of fuel assemblies. The increase of interval periods between refuelling outages reduces personnel exposure. Experimental fuel assemblies built by FRAGEMA are currently being irradiated to assess the high burnup performance in commercial PWRs.

4.1.2 Primary System Material

Considering the materials not directly submitted to neutron flux, the main contribution of activated corrosion products is by steam generator Inconel tubing, which represents 75 % of PWR surfaces cooled by the primary coolant. The choice of materials (Inconel 690) is governed by the corrosion resistance behaviour in primary and secondary chemistries.

Cobalt is also present in stellite, which is commonly used for hard-faced coatings (control rod drive mechanism (CRDM), valves, centring keys, etc ...).

Wear and erosion of hard-faced materials and release of cobalt particles in the primary system contribute heavily to an increased dose field. FRAMATOME has undertaken tests to qualify alternative materials to stellites in order to replace the cobalt contained in stellite by either cobalt-free, hard-faced materials such as colmonoy or cenium alloy, or by thin, hard-faced coatings with a low cobalt content.

Endurance testing of CRDM equipped with coated latch arms with low cobalt material on a loop simulating PWR conditions indicate a drastic reduction of wear rates and consequently a significant reduction of cobalt particles within the primary coolant.

The quality of surface conditions of primary components is of a prime importance for dose field reduction.

A joint experiment between EdF, CEA and FRAMATOME carried out on mainhole flange gaskets of steam generator channel heads, shows a drastic reduction of the surface activity of 4 between prepolished conditions (mechanical + electropolishing) and the as grinded conditions after the 1st cycle of operation of the Chinon Plant.

4.2 Decontamination

Decontamination is currently being considered and used to facilitate steam generator replacement.

Two methods of decontamination processes are implemented to minimize the personnel dose exposure during steam generator replacement operation :

(1) Electrodecontamination of the reactor coolant piping ends after SG removal. Electrodecontamination is carried out by means of a set of two leaktight suction pads applied to the reactor coolant piping walls and supported and driven by a carrier.

The carrier drives both the continuous forward motion and rotation of the suction pads in which the circulation of the electrolyte (PO_4H_3) and the current density are adjusted according to specified values to obtain the appropriate decontamination factors.

(2) Dilute chemical decontamination process based on the LOMI technique from which FRAMATOME acquired the license in 1987. In this process decontamination is performed by circulating an alternatively oxidating and reducing solution by the means of injection lance (see fig. 5).

In the case of Dampierre 1 steam generator replacement, the results obtained by using both decontamination methods indicated a reduction of the initial contamination level of 13. In addition the total personnel doses for the Dampierre steam generator replacement operation remain below 220 man rem or 2.2 man Sievert.

4.3 Development of remote controlled devices

Maintenance on reactor equipment is carried out during plant shutdown. To reduce the ORE of maintenance personnel, the FRAMATOME Service Division has developed and built a large variety of remote-controlled tooling in order to avoid the presence of maintenance personnel in high radiation area.

Steam generators, which are part of the reactor coolant system, require various regular inspection and maintenance operations covering primary and secondary sides.

Several remote-controlled devices have been built to perform periodical inspection and maintenance of steam generator tube bundles, eg

(a) The torquing machine for folding covers, which allows isolation of the steam generator channel head from the reactor coolant inlet and outlet nozzles, and allows the operator to monitor each operation away from high radiation areas.

(b) The maintenance spider, with its servicing equipment which can be installed and fixed on the steam generator tube sheet from the outside. This fully automated maintenance spider travels over the surface of the tube sheet and can reach its working position automatically ; it can be equipped with a wide variety of tools such as the milling machine, welding torch, obturation plugs, eddy current probes, cameras. This equipment is of immeasurable benefit to maintenance because personnel are no longer needed in high radiation areas.

(c) Various servicing arms for performing dedicated repairs inside the steam generator channel head such as : tube plugging, tube end machining, tube extraction ...

(d) A remote-controlled mobile manipulator is currently under development to perform large steam generator repairs inside the steam generator channel head. This manipulator is designed to have a stronger load capacity than the maintenance spider and to extend the possibility of repairs without the presence of personnel in high radiation areas.

5 CONCLUSION

An extensive effort has been devoted by EdF plant operator and owner and by FRAMATOME plant NSSS designer and constructor for reducing Occupational Radiation Exposure (ORE) and radwaste volumes.

An on going R&D effort is directed toward a better volume reduction and waste immobilization of effluents produced by operating plant.

After some years of operation, results obtained show a drastic reduction of Dry Active Waste Volumes : 180 m^3/plant and a trend of low Occupational Radiation Exposure (ORE) of plant personnel : 180 man rem/plant and low (ORE) versus plant electric generation 35 man rem per billion of KWh.

WASTE TREATMENT OVERVIEW

FIG : 1

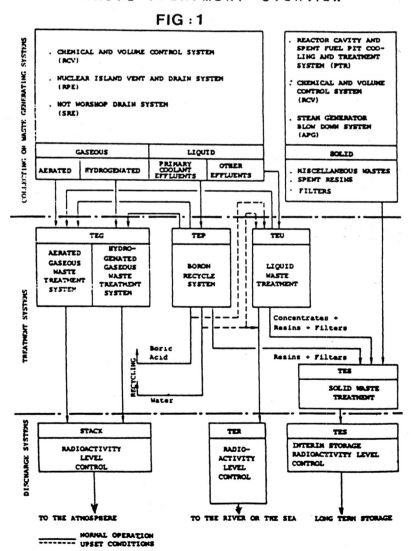

FIG :2

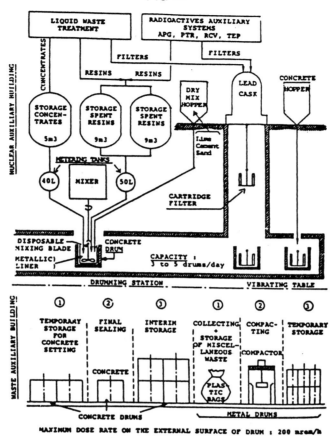

**Volume des déchets
radioactifs conditionnés**

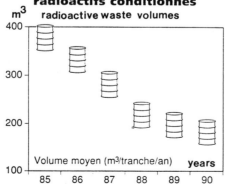

radioactive waste volumes

*Après une baisse très forte,
le volume des déchets continue de décroître.*

FIG 3

★ COLLECTIVE DOSE DATA 1988 ★

	NUMBER OF UNITS IN OPERATION	ANNUAL DOSE PER REACTOR IN MAN REM	DOSE PER TWH GENERATED OR MAN REM PER BILLION OF KWH
US BWR	36	511	117
US PWR	70	345	62
JAPANESE LWR	35	270	54
KWU PWR	12	260	38
FRAMATOME PWR	50	180	33

106

FIG : 4

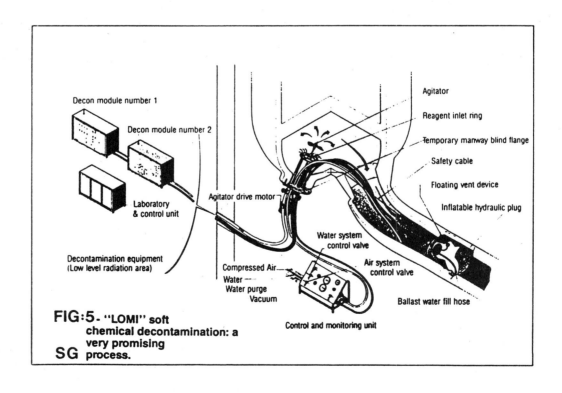

FIG:5. "LOMI" soft chemical decontamination: a very promising SG process.

C431/076

Trends in American practice in waste volume reduction

G C LILLY
Container Products Corporation, Wilmington, North Carolina, United States of America
C J F LETTINGTON, BSc, PhD
GEC Alsthom Engineering Systems Limited, Whetstone, Leicestershire

SYNOPSIS This paper describes the approach that has been adopted by the American low level active waste industry and the US utilities to minimise both the waste volumes produced by the industry and disposal costs. Decontamination with minimal waste arisings, waste sorting, compaction up to an economically determined limit and packaging are briefly reviewed.

1 INTRODUCTION

Volume reduction of low level radioactive waste (LLRW) has become increasingly important in the USA over the past several years due to increasing disposal costs and impending changes regarding disposal site locations, resulting in the anticipated need for interim storage. In the five year period of 1985-1989 the total volume of waste received at commercial burial sites decreased by 39% per annum. The LLRW received for disposal which had been generated by the nuclear power industry during 1990 was an estimated 22,680 m³, (an approximate reduction of 5% from 1989) which translates to about 347m³ of LLRW generated at a typical 1000 MWe light water reactor per annum.

A number of innovative procedures and new equipment applications have proved to be successful in significantly reducing the volume of waste requiring disposal and are thus contributing to this improved situation.

The three main steps are: minimisation of the amount of material which becomes contaminated, waste material sorting to separate items that can be economically decontaminated and compaction of the LLRW to reduce its volume. These can be linked into an integrated system as shown in Figure 1.

2 DECONTAMINATION

One significant source of waste is the protective clothing required for personnel access to contaminated work areas. Efficient decontamination of these areas obviously reduces this source of waste, but the method used for decontamination must be selected with particular attention to the amount of secondary waste generated during the procedure.

The "Kelly" decontamination system provides an effective method for decontaminating general work areas. This method, which began commercial operation in 1982 is now widely used in the USA. The basic technique is to force water at a temperature of typically 135°C on to a contaminated surface and into its fissures.

The water flashes to steam and so creates a powerful drag force on the contaminated particles away from the surface. The water, steam and contaminants are immediately removed by a vacuum system. The secondary solid waste generated by this method of decontamination is found to be as little as 10% or less of that generated by other methods. The secondary liquid waste generated is approximately 10% of that volume produced by other methods. Furthermore, the reduced volumes of secondary waste generated are a recognised bonus to the effort of decontaminating an area to the level at which personnel are allowed access without protective clothing, thus eliminating a major source of low level radioactive waste.

The same technique can be used to decontaminate chains, tools, motors and other items of equipment by spraying water at about 135°C on to these items whilst they are contained within a glove booth or other modular enclosure. Figure 1 includes the use of a glovebox for this purpose. Steam and air are educted through a demister and HEPA filter and the contaminated liquid waste is normally fed to the station's active drains system. Incoming air is routed across the windows to maintain visibility.

3 WASTE SORTING

Effective management and planning reduces the amount of waste generated and segregates the waste by form and activity in preparation for ultimate disposal. A final sorting operation prior to packaging provides the opportunity to reduce the amount of material requiring disposal under the regulations applicable to radioactive waste. A sorting operation provides an opportunity for operators to separate out materials, or items of equipment, from the waste stream which can be readily decontaminated and reused. It is also practical to separate less radioactive materials which can be disposed of under less stringent regulations, either with or without decontamination.

Sorting systems are commercially available in the USA which allow radioactive waste to be safely sorted by an operator who is not wearing protective clothing. They consist of totally enclosed conveyors passing through a series of glove ported negative pressure sorting stations. Material which is not appropriate for decontamination can be uniformly distributed within a waste container. Such a system can be operated in isolation, but they are often linked directly to a compactor.

4 COMPACTION

Once all practical effort has been expended to reduce the volume of waste requiring disposal, the most important consideration is the selection of the packaging for the LLRW which will provide optimum use of disposal space and minimum labour costs while complying with the applicable regulatory requirements. Rectangular steel containers can be designed to meet these requirements much more effectively than drums, which inevitably contain voids when they are stacked together. The rectangular steel container, marketed in the USA as the model B25, provides the volume otherwise available by using twelve 208 litre (55 US gallons) drums, but only requires the floor space of six drums. Significant savings in labour and transportation also result from the use of the larger rectangular containers. Rectangular containers of other dimensions are specified by some utilities with comparatively small changes to the advantages that they enjoy over drums.

A container design can be modified to meet various applications and regulatory requirements based on variables such as isotopic content, specific activity, weight and physical configuration of the material being packaged. In addition to the reduced space requirements for storage and transportation, utilisation of the model B25 results in significant labour savings. One reason for this is that the packaging of larger more complex items is possible without requiring the size reduction which is often associated with the use of

relatively small cylindrical drums. Labour savings are further realised, over the use of drums, by reducing from 12 to 1 the number of filling, closing, labelling and manifesting operations required for the same quantity of waste.

The use of compaction in the USA to reduce waste volumes is extremely common and compaction pressures have in the past risen steadily. However, as the LLRW becomes more compact, i.e. there is less air in the drum or rectangular container, further volume reduction becomes more difficult to attain. A compaction pressure can therefore be reached after which a further incremental investment in a more powerful compactor does not yield a satisfactory saving in the costs of waste disposal, handling and transportation.

A representative mixture of LLRW in the USA is given below.

Table 1. Material Mix of LLRW

Material	% by Weight
plastic – sheets, rolls etc	25
pvc tubing	10
cloth	10
paper	25
metal, pipe, rods, tools etc	15
soil	5
composite materials	10

On the basis of the material mix given in Table 1, Fig 2 has been developed which shows the density increase versus the compaction pressure used. The decline in the rate at which the density increases with increasing pressure is marked above compaction pressures of around $1000kN/m^2$.

These diminishing returns have been taken into account by the author and his colleagues in the design of a compactor which is marketed as the B400 Supercompactor. This machine is designed to compact materials in model B25 containers incorporating a system for maintaining compression forces on material that has already been compacted whilst more material is fed into the container. The compactor is designed to exert a minimum force of 1.78MN (400,000 pounds) over an area of $1.84m^2$ through a compaction ram matched to the internal dimensions of the container. This force is generated by a hydraulic system driving two vertical cylinders through a control unit which is based on a programmable logic controller, programmed to sense areas offering the greatest resistance to the compaction force. Through mechanical compensation, the force is increased in these areas of resistance far in excess of 400,000 pounds. During the operating cycle the compactor continuously senses this localised resistance and automatically increases the compaction force accordingly to provide optimum compaction.

© IMechE 1991 C431/076

The system typically produces volume reduction factors of 10 or greater when compacting a waste material mix such as that commonly produced in a nuclear power generating facility.

Another advantage of compacting material directly into rectangular metal packaging with volume reduction factors of 10 or greater is the cost benefit over systems which compact materials and packaging together into smaller shapes. In the USA these systems produce either rectangular or cylindrical "pucks" which require overpacking prior to disposal. The additional handling and packaging adversely affects both the overall compaction achievable and the labour costs.

5 CONCLUSIONS

In summary, the approach to volume reduction of LLRW in the USA is the management of all activities which affect the low level radioactive waste stream. This effort begins with waste minimisation procedures to generate the least possible amount of waste followed by the assurance that material which is reusable or disposable under alternative procedures is removed from the radioactive waste stream. Finally and possibly the most important factor in terms of the impact on a station's operating costs, is the packing and compaction of the radioactive waste.

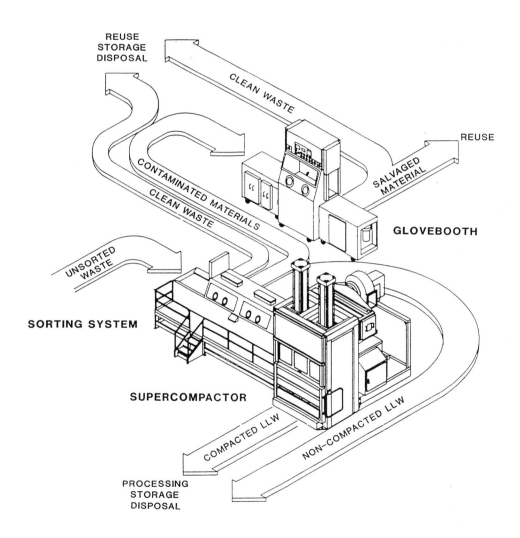

Fig.1 An Integrated Volume Reduction System

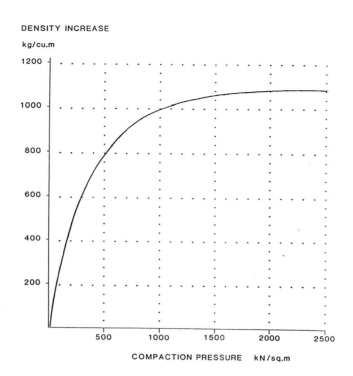

Fig.2 Effect of Compaction Pressure
on LLRW Volume

The modified MOWA, the transportable radioactive waste conditioning plant

J SIMS, MINucE
Wastechem Limited, Bramhall, Cheshire

SYNOPSIS

MOWA, the MObile WAste Conditioning Plant, was originally designed to encapsulate a variety of liquid radioactive wastes in a cement matrix, mixing the products in the final disposal drum employing the lost paddle principle. Following operational experience gained over a seven year period, MOWA has been modified to eliminate the in-drum mixing paddle, thus reducing the mixing power requirements, removing the upper capacity limit on the final disposal package and minimising the radiation dose uptake of the operational staff.

1.0 THE ORIGINAL MOWA

1.1 MOWA was designed and constructed in the early 1980's with the prime objective of being sufficiently portable to permit it to visit a variety of nuclear sites to encapsulate β / γ emitting radioactive wastes such as sludges, ion exchange materials, concentrates, etc. During its operational history, MOWA undertook 38 different encapsulation campaigns treating over 1,100 M³ of radioactive material with just one plant.

1.2 This transportable conditioning plant had the following technical specification:

Table 1

Overall dimensions dimensions	5.7 metres(L) x 2.22 metres(W) x 2.28 metres(H)
Weight	approx 22,000 Kg
Throughput	Sludge up to 10 m³ per 8 hr shift Granular resin up to 2m³ per 8 hr shift
Controls	Fully programmable for automatic operations
Electrical Supply	2 x 63A 380V
Utilities	Water @ 6 bar
Setting up time	1 day
Operational Staff	1 supervisor, 2/3 operatives

1.3 The standard MOWA, which is illustrated in Figure 1, consisted of two independent dosing and mixing positions for the waste disposal drums, each of which contained the correct quantity of cement (or ester vinyl) binder and a lost paddle. Also mounted on the basic structure was an accessible, but sealed, ventilated cell containing two metering vessels, one for sludges and one for ion exchange resins and all of the associated pipework, valves and pumps, etc. The metering vessel for the ion exchange resin was fitted with a sieve to permit a degree of dewatering of the resin to be achieved. The MOWA assembly was completed with two hydraulic power packs for the drives of the in-drum paddles and an air compressor and reservoir to control the valve actuators.

Figure 1   The Standard MOWA

1.4 The plant was equipped with a fully programmable control system for the automatic operation of the process, and a portable control console.

1.5 Heavy lead shielding installed on or around all equipment carrying radioactive material, that is, the metering vessels, pumps, pipework and valves, and steel shielding at the mixing position, in conjunction with disposal drums incorporating radiological protection, reduced the radiation exposure to 100 micro Sieverts/hour at the surface of the plant while treating waste with a specific activity of 37 TBq/m^3 (Co60) equivalent).

1.6 The process diagram of the original MOWA, clearly indicating the sealed and ventilated containment cell is shown in Figure 2.

1.7 Prior to putting MOWA to work on any site, a non-radioactive simulate was produced based upon a chemical analysis provided by the site owners. This simulate was used to establish the correct formulation in a cementation laboratory equipped with an exact replica of the MOWA mixing system, the formulation being approved by the disposal authorities, before the commencement of the campaign.

1.8 MOWA was also approved by the licensing authorities following an extensive review of the safety aspects of the plant.

1.9 The original MOWA was purchased by the following organisations:

Transnuklear, Germany (1978)
Nucleco, Italy (1982)
Studsvik, Sweden (1983)
Paks, Hungary (1986)

2.0 THE MODIFIED MOWA

2.1 A number of ad-hoc approved modifications were made to the MOWA during some of the later campaigns, all of which proved very successful and were the basis of more permanent changes leading to the plant being called MOWA II.

2.2 It will be seen from the process diagram of the modified plant, Figure 3, that the contained, ventilated metering tanks, pipework, pumps and valves are unchanged. The principal changes have occurred at the drum mixing stations which have been deleted, as have the hydraulic power packs.

2.3 The in-drum mixers have been replaced with a single batch mixing tank of 400 litre capacity, a heavy duty peristaltic pump which has its inlet connected to the bottom of the batch mixing tank, and a cement weighing bin. In addition, a cement storage hopper is also required by the MOWA II, but this can be installed outside the radiological area. The storage hopper is connected to the weighing bin by means of a pneumatic cement transport system. The weighing bin is connected to the batch mixing tank by a second pneumatic transfer system.

2.4 The output from the peristaltic pump is fitted with a flexible pipe delivering the products of the batch mixing tank to a waiting disposal drum of any chosen capacity.

2.5 In addition to the lead shielding installed around the radioactive waste carrying equipment in the sealed cell, the batch mixing tank, peristaltic pump and flexible delivery pipe are all supplied with suitable lead shielding.

2.6 Because all motors on the MOWA II are electrically driven, the two heavy hydraulic power packs at the rear of the plant have been deleted. Although the batch mixing tank, peristaltic pump, radiological shielding and weigh bin occupy more space than the original drum mixing stations, the savings made by deleting the hydraulic power packs means that the overall length and weight of the MOWA has remained unchanged (see Table 1).

2.7 The MOWA II is, like its predecessor, provided with a fully programmable control system for automatic operation and a portable control console.

2.8 The layout of the plant is shown in Figure 4.

3.0 THE MOWA II PROCESS

3.1 As has been discussed in para 1.7, it is first necessary to establish the correct formulation for encapsulating a site's waste in cement, although due to the variable speed capability on the batch mixing tank, an ester vinyl matrix could be used if necessary. In the latter case, a chemical dosing plant would be required alongside the MOWA II.

3.2 It may be necessary at an early stage to select the final container size, particularly if special units larger than the readily available 200 and 400 litre capacity units are to be employed.

3.3 The plant has to be installed within a radiologically controlled building, preferably with its own ventilation system. This building will also have been supplied with adequate connections to the sites' waste storage/pumping systems via ball valve couplings.

3.4 On delivery to the site, the MOWA II is connected to the appropriate electrical and radioactive waste systems, and on demand, the metering vessels will be filled with the appropriate waste material.

3.5 At the correct time in the automatic control sequence, the contents of the metering vessels will be transferred to the batch mixing tank by the pneumatic double diaphragm pump. Because the ion exchange material metering vessel is fitted with a sieve, some dewatering can be achieved by pumping away the free water below the sieve unit prior to transferring the material to the batch mixing tank.

© IMechE 1991 C431/003

3.6 Additional dewatering of ion exchange material can be achieved if required, via the dip leg in the batch mixing tank prior to the cement being added from the weigh bin.

3.7 Following completion of the mixing process, the products of the batch mixing tank are transferred to the waiting disposal container by means of the peristaltic pump.

3.8 This process will be continued until the disposal container has been filled with encapsulated material.

3.9 Finally, the peristaltic pump and shielded pipework to the disposal container are cleaned by placing a number of foam rubber balls into the system via a special connection. These contaminated balls form part of the last mix entering the final disposal container.

4.0 ADVANTAGES OF THE MOWA II CONCEPT

4.1 When using the original MOWA plant it had proved necessary to encapsulate a high volume of water, which was required to achieve efficient pumping of the ion exchange material, particularly when cement powder had been previously dosed into the disposal drums, and where it was impossible to achieve any additional dewatering. The revised arrangement of filling the batch mixing tank with waste, prior to adding the cement means that extra dewatering can be undertaken in the tank after pumping has been terminated, but prior to the arrival of the cement powder.

4.2 Adding the cement powder to the wet waste means that the resultant torque mixing curve is much more favourable than with the original MOWA, when the system of adding wet waste to the cement already in the mixing drum produced a very severe torque requirement, whilst at the same time requiring a powerful restraint to prevent the mixing drum rotating under that torque. Both high mixing torque and drum restraint requirements have been eliminated.

4.3 The design of the original MOWA restricted the size of the drums which could be fitted under the mixing heads to either 400 litre unshielded or 200 litre shielded. Although MOWA II can still fill such drum sizes if required by a specific client, it is totally unrestricted on the maximum capacity of the disposal container due to an ability to fill the container in a number of batches.

4.4 The cost of the lost paddle in each mixing drum has been eliminated by undertaking the mixing of all of the encapsulation material in the batch tank prior to transferring those mixed products into the final disposal container. Such a process

also means that large solid decommissioned equipment could be placed into the disposal container, and flooded with the products of the batch tank, an operation originally impossible due to the need to rotate a mixing paddle in the final drums.

4.5 As fewer but larger disposal containers need to be manhandled after filling the operator's radiation dose uptake is considerably reduced over a normal operating period.

REFERENCES

(1) BRUNNER H, CHRIST B, and KOHLPOT W – Solidification of liquid wastes from nuclear power plant by mobile solidification units. Symposium on the on-site management of power reactor wastes, Zurich, Switzerland, 1979.

(2) CHRIST B, VYGEN P – Operating Experience with high standard mobile solidification units. ANS topical meeting, Waste Management, Tucson, 1982.

(3) FILTER H E – Utilising existing technology to improve environmental protection. ANS annual meeting, Detroit, Michigan, 1983.

(4) BRUNNER H, GANSER B, MANTOVANI I M and ROYSTON J – Eine mobile Anlage zur Verfestigung radioaktiver Abfälle mit dem Dow-Binder 101. Jahrestagung Kerntechnik, Frankfurt, 1984.

(5) BRUNNER H, GANSER B, KLEIN G – MOWA the mobile waste conditioning plant. Conference on Radioactive Waste Management, BNES, London, 1984.

(6) SIMS J – The Operation of a Transportable Radioactive Waste Solidification Plant, Nuclear Engineer, Volume 29 No 6, November 1988.

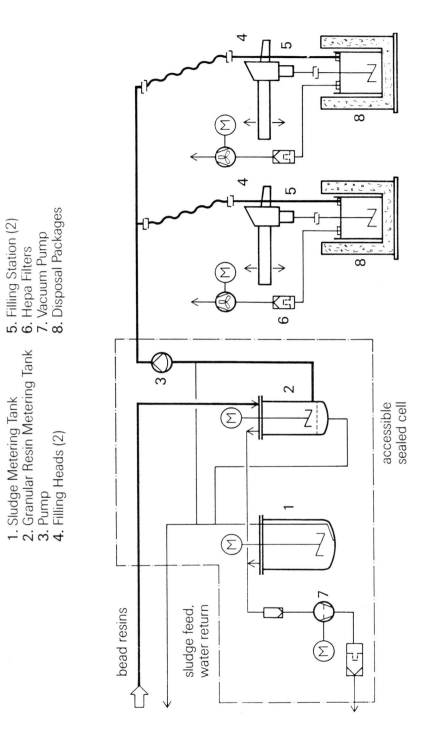

1. Sludge Metering Tank
2. Granular Resin Metering Tank
3. Pump
4. Filling Heads (2)

5. Filling Station (2)
6. Hepa Filters
7. Vacuum Pump
8. Disposal Packages

bead resins

sludge feed,
water return

accessible
sealed cell

Fig 2   Simplified Process Diagram of MOWA

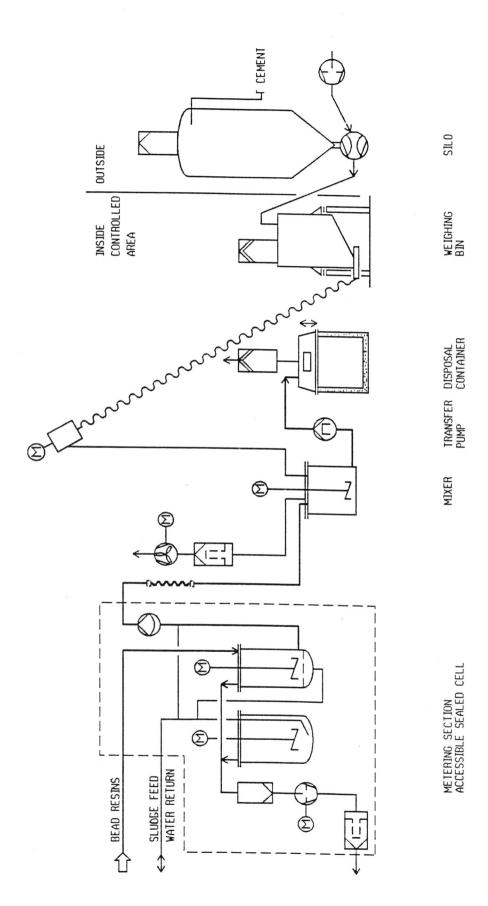

Fig 3 Simplified Process Diagram of MOWA II

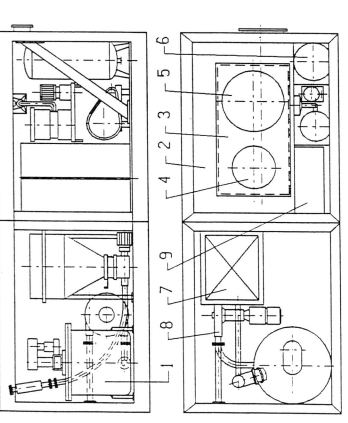

MOBILE FACILITY

1 MIXER
2 SEALED CELL
3 TANK SHIELDING
4 METERING TANK FOR RESINS
5 METERING TANK FOR CONCENTRATES
6 AIR COMPRESSOR
7 WEIGHING BIN
8 TRANSFER PUMP
9 CONTROL AND SWITCH CABINET

TECHNICAL DATA

OVERALL	LENGTH	5600MM
DIMENSIONS	WIDTH	2200MM
	HEIGHT	2200MM
WEIGHT	APPROX.	22000KG

Fig 4 Mobile Waste Conditioning Facility

C431/023

The development of radiometric waste monitoring systems at the Sellafield reprocessing complex

N GARDNER, PhD, J C B SIMPSON, PhD, CPhys, MInstP, G H FOX, BSc and A McGOFF, BSc
British Nuclear Fuels plc, Sellafield, Seascale, Cumbria

SYNOPSIS The safe handling, routing and disposal of waste at the Sellafield reprocessing site requires the provision of monitoring instrumentation capable of detecting and quantifying the activities of specific radioactive isotopes and/or its fissile content. Wastes may contain a variety of matrix materials and activity levels varying from <0.4Bq/g for Very Low Level Waste, to >10^{10} Bq/g in residual fuel bearing wastes. The waste may be contained in small packages or very large crates. BNFL have developed monitoring instruments where no commercial systems are available, the emphasis being on the cost effective achievement of accurate and reliable measurement.

The combined efforts of the Sellafield Technical Department and the Systems and Equipment Engineering Department at Risley have ensured the successful completion of many projects. These have ranged from large, complex instruments installed in fuel handling and chemical plants, where they form an integral component of the overall plant control system, to small free standing and independent devices for the rapid preliminary sorting of waste at source. Measurement techniques embodied in these systems have been equally varied, ranging from those based on conventional gamma spectrometry and total neutron counting to coincidence and interrogation techniques. These radiometric techniques are used both singly or in combination and are linked, where necessary, to predictive algorithms for radioisotopes which are not directly detectable.

This paper describes the methodology of instrument development which has been adopted to ensure that customer requirements are met, gives examples of systems which have been developed or are under development and identifies some areas of development in support of future requirements. Examples given include the quantification of the full inventory of radioactive isotopes in the intermediate level waste stream arising from Magnox fuel decanning, the measurement of fissile material in miscellaneous beta gamma waste and an ongoing study of Very Low Level Waste discrimination systems.

1 INTRODUCTION

The safe handling, temporary storage and ultimate disposal of radioactive waste must be carried out within a framework of national regulations and criteria; to demonstrate compliance requires the provision of appropriate radiometric systems. Although at present not all disposal criteria are specified, it seems very likely that a detailed radioisotopic inventory will be required for all waste products before they are accepted for disposal. In some cases it may be possible to provide satisfactory information on the basis of known parameters relating to the waste stream whereas, in others, radiometric measurements may be the primary, or only, source of information.

The Physical Science and Engineering Development (PS & ED) group in the Sellafield Technical Department (STD) has extensive experience of the development of specialist instrumentation to satisfy the monitoring requirements of current and future reprocessing, waste management and decommissioning operations at Sellafield.

Work currently being carried out by PS & ED group relating to the field of waste measurement includes the development of techniques and monitoring systems for the following:-

(a) In-plant inventory measurements on wastes as they arise from front end operations in both the Fuel Handling Plant (FHP), dealing with reprocessing of Magnox fuel, and the Thermal Oxide Reprocessing Plant (THORP), dealing with uranium oxide fuel from a variety of sources. In the case of FHP, the Swarf Inventory Monitor (SIM), which is now operational, provides on-line inventory information on Magnox swarf arising from decanning operations prior to transferal to Encapsulation Plant 1 (EP1). The Hulls Monitor, currently under construction in THORP, will provide similar information for waste transferred to Encapsulation Plant 2 (EP2).

(b) Measurements to determine the inventory of Low Level Waste (LLW) consigned to the Company's Drigg disposal site, from all sources, to verify compliance with the Drigg Conditions of Acceptance. These will be carried out in the new Waste Monitoring and Compaction (WAMAC) facility.

(c) Measurements on Intermediate Level Waste (ILW) including Plutonium Contaminated Material (PCM) arising from past, current and future decommissioning operations on the Sellafield site. Since a suitable depository for ILW does not yet exist, the inventory requirements are not yet fully defined. Hence, present work is being carried out in anticipation of the likely NIREX disposal requirements.

(d) Measurements on ILW for interim storage in the Miscellaneous Beta/Gamma Waste Store (MBGWS), prior to disposal in a future depository. The fissile content of waste sent to this facility is determined, in order to demonstrate compliance with criticality clearance conditions, in an integral facility known as the Fissile Material Detector (FMD).

(e) Measurements on decommissioning waste arising at the LLW/VLLW (Very Low Level Waste) boundary in order to sentence for Drigg disposal or for reuse/'free-release' as appropriate.

All the measurements described above present individual problems for which there are no common solutions. Some of the requirements will be impossible to satisfy whereas, in other cases, a solution to a specific problem may be applicable in another area. In all cases, the instrumentation produced must be extremely reliable and, in the unlikely event of failure, be failsafe. This paper briefly describes the methodology which is used in the development of these radiometric waste monitoring systems, gives a number of specific examples of systems which have been developed or are under development, and identifies some of the areas in which development work is being undertaken.

2   METHODOLOGY OF INSTRUMENT DESIGN

In order to ensure that customer requirements are met in full, a detailed Measurement Specification (MS) is prepared in consultation with all interested parties. The MS details the measurement(s) required at a particular point in the process together with other relevant information, such as background information relating to the plant, reasons for the measurement(s), measurement conditions and constraints, system control and interfaces, performance required (measurement accuracy and precision, reliability and lifespan) and project timescales. The merits of all applicable measurement techniques are then considered in a preliminary feasibility study which normally involves both theoretical and experimental investigations. Usually, the technique utilised is based on one or a combination of the following:-

Gamma Spectrometry: both High and Low Resolution (HRGS and LRGS) (1,2).
Gamma-ray and X-ray Absorptiometry: involving radiation attenuation in the material under investigation (2).

Passive Neutron Counting: both Total Neutron Counting (TNC) and Neutron Coincidence Counting (NCC) (2).
Neutron Absorptiometry: involving the attenuation of neutrons by the material under investigation (2).
Neutron Interrogation: eg. via a Neutron Generator in the Differential Die-Away (DDA) technique (3).

Other techniques are sometimes used when appropriate, for example X-ray spectrometry. In cases where it is not possible to directly determine the activity of a particular radionuclide, it is necessary to carry out the determination based on inference from the measured activities of other radionuclides. This usually involves characterisation of the waste stream using computer models, for example, the FISPIN code (4) for determining activity inventories of fuel as a function of initial enrichment, burn-up and cooling time.

Following acceptance by the customer of the recommendations of the initial feasibility study, a decision is made on the most appropriate route and a detailed System Specification produced in parallel with the finalisation of the main design parameters. The detailed design itself is then produced based on the information contained in the System Specification. This is followed by equipment procurement in combination with finalisation of the installation and commissioning timetables. Subsequent to commissioning of the instrument there is usually a period of operational support after which the operators manual and a final version of the System Specification are issued, taking account of any design changes resulting from the installation and commissioning. On completion of a training period for the plant operators the system is formally handed over.

The following section gives some examples of radiometric waste monitoring systems that have been, or are in the process of being, developed by BNFL using this methodology.

3   EXAMPLES OF WASTE MONITORING SYSTEMS AT SELLAFIELD

3.1 _Swarf Inventory Monitor_

Magnox swarf from decanning operations in the Fuel Handling Plant (FHP) was originally stored in silos. Since July 1990 this swarf has been processed in the new waste Encapsulation Plant (EP1). PS & ED group were charged with the task of providing instrumentation to satisfy a process control requirement to minimise fuel losses from the plant and, in order to satisfy the anticipated NII and DoE criteria for long term storage/ disposal of waste, to provide a full radio-active inventory of the waste. Therefore the so-called 'Swarf Inventory Monitor (SIM)' had to satisfy two basic measurement requirements which were identified in the Measurement Specification:-

(i) To estimate the mass of uranium in magnox swarf on the swarf sorting trays in the magnox decanners, and to alarm if this exceeds a set value. Each measurement cycle involves monitoring swarf and associated fuel residues from up to 4 fuel rods, with measurements required at a frequency of up to 20 per hour.

(ii) To determine the activity inventories for 44 specified radioisotopes (see Tables 1a and 1b) for individual bins of magnox swarf exported from the decanners to EP1.

Table 1a Isotope activities directly determined by SIM

Isotope	Half Life (Years)	Isotope	Half Life (Years)
Co-60	5.27	Ag-110m	0.68
Zn-65	0.67	Cs-134	2.065
Zr-95	0.17	Cs-137\	30.17
Nb-95	0.10	Ba-137/	
Ru-103\	0.11	Ce-144\	0.78
Rh-103/		Pr-144/	
Ru-106\	1.02	Eu-154	8.59
Rh-106/			

Table 1b Isotope activities calculated by SIM

Isotope	Half Life (Years)	Isotope	Half Life (Years)
C-14	5.7×10^3	U-234	2.5×10^5
Se-79	6.5×10^4	U-235	7.0×10^8
Sr-90	28.5	U-236	2.34×10^7
Zr-93	1.5×10^6	U-238	4.47×10^9
Tc-99	2.13×10^5	Np-237	2.1×10^6
Ag-108m	127.	Pu-238	87.7
I-129	1.6×10^7	Pu-239	2.4×10^4
Sb-125	2.76	Pu-240	6.6×10^3
Sn-126	1.0×10^5	Pu-241	14.4
Cd-133m	13.7	Pu-242	3.8×10^5
Cs-135	3.0×10^6	Cm-244	18.11
Pm-147	2.62	Am-241	433.
Sm-151	88.7	Am-242m	141.
Eu-152	13.3	Am-243	7.4×10^3
Eu-155	4.96		

In order to fulfil the measurement requirements, a system based on the use of High Resolution Gamma Spectrometry (HRGS), in combination with inferential techniques, was chosen. Measurements are carried out on small batches of swarf prior to its placement in the swarf bin for transport to EP1. (Each EP1 product contains the swarf from one swarf bin). A schematic diagram of the specified monitoring arrangement is shown in Figure 1.

The bulk of the development programme in support of the System Specifications for this instrument was related to the data analysis algorithms. Fuel irradiation and cooling time are determined via isotopic activity ratios which were developed for a predecessor of SIM (5).

The fuel mass for process control purposes is determined from the Cs-137 activity and the Cs-137 activity/gram obtained from the

irradiation determination. The algorithms include a correction for gamma absorption within the fuel debris based on the ratios of gammas from a single isotope. This correction is extremely important to the accuracy of the result as the residual fuel can vary in physical form from 'fines', ie. very small particles, to complete sections of fuel rod. The calculated fuel mass is then transmitted to the FHP decanner control computer.

Fig. 1 Schematic layout of the Swarf Inventory Monitor in the Fuel Handling Plant Decanner Cell

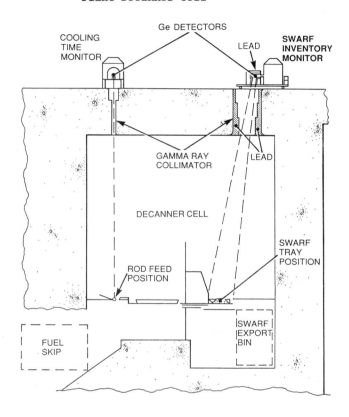

The activity inventory is derived from a direct measurement of the detectable gamma emitters listed in Table 1a and by inferring the activities of the remaining isotopes listed in Table 1b. In the direct inventory measurements a more sophisticated 'iterative' absorption correction technique is used than in the process control mass determination - this allows for the presence of a mixture of fines and large pieces of fuel. The inferred isotopic activities are determined using data from the computer code FISPIN, the measured Cs-137 activity and the irradiation and cooling time data. The total inventory is then transmitted to EP1.

Figure 2 is a schematic diagram showing the general arrangement of the electronics hardware. Whenever possible commercial electronics and software has been used, but a considerable amount of in-house software has been written. Where necessary commercial equipment has been modified to suit BNF standards. In particular the germanium detectors have been given greatly improved immunity to RF interference. Because of the high specific activity of the waste stream and the requirement for rapid high sensitivity measurements, the system had to be capable of handling a range of input count rates up to 500,000 per sec. Hence, development work was carried out in order to maximise the signal throughput of the electronics and to minimise errors resulting from operating with high system deadtimes.

Fig. 2 Schematic of the electronics hardware of the Swarf Inventory Monitor

(DDA) technique was chosen. This involves the injection of short pulses (a few microseconds) of fast neutrons, from a neutron generator, into a measurement chamber constructed from moderating materials (Figure 3). The fast neutrons are slowed down by the moderating materials to produce a thermal neutron flux within the assay chamber which persists for a few milliseconds. These

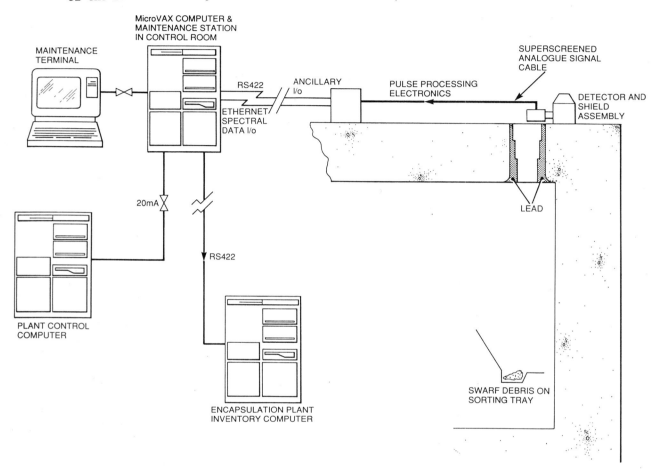

The monitor has been operating reliably since July 1990. Results have been checked using fuel 'pennies' of known masses cut from rods of known irradiation and cooling time. In these checks the mass derived from the process control measurement is compared with the mass determined from the inventory calculation and both compared with the fuel penny mass. Agreement is found to be good. Inventory measurements have been checked by comparing directly measured activities with the inferred activity of the same isotope. Again agreement was found to be good.

3.2 Fissile Material Detector

The measurement requirement of the FMD is to measure the total fissile content of packages of miscellaneous beta/gamma waste, as they are consigned to the MBGWS, and to provide an alarm/trip output if the fissile content exceeds 10g Pu-239 equivalent. This trip output is then used to inhibit the movement of the package into the storage box.

In order to satisfy the measurement requirement, the Differential Die-Away

thermal neutrons induce fission events in any fissile material present in the assay chamber, thereby producing additional fast neutrons. A typical fast neutron detector response as a function of time for the FMD DDA facility is shown in Figure 4, this shows the very high initial count rate, due to the interrogating neutron pulse, followed by a more slowly decaying signal due to neutrons arising from fission events. It is these fission neutrons that are counted in order to provide a measure of the fissile content of the waste package.

The FMD has been calibrated to deal with the wide range of wastes and waste containers that are consigned to the store. The approach taken was to ensure that the systematic error associated with a measurement would never allow the value of fissile material (mass + error) to be underestimated, thereby ensuring criticality safety. The systematic error arises from both the geometrical variation in response within a single waste matrix and the variation in mean response between different wastes falling within the same category. The values of the systematic errors to be applied were derived during calibration and are dependent both on the type of waste (filters,

Fig. 3 Schematic of the Fissile Material Detector DDA Facility

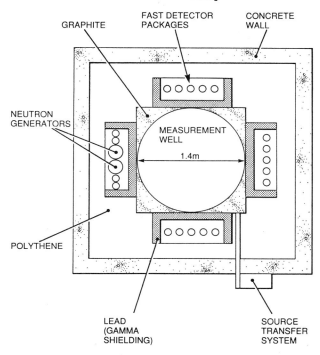

Fig. 4 Typical response curve for neutron detection during a single operation cycle of a DDA facility

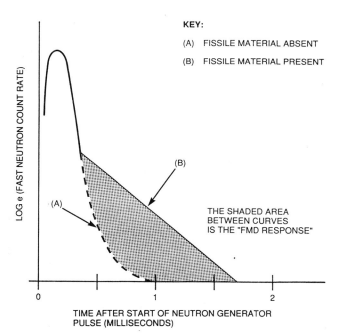

KEY:

(A) FISSILE MATERIAL ABSENT

(B) FISSILE MATERIAL PRESENT

THE SHADED AREA BETWEEN CURVES IS THE "FMD RESPONSE"

LOG e (FAST NEUTRON COUNT RATE)

TIME AFTER START OF NEUTRON GENERATOR PULSE (MILLISECONDS)

combustible and non-combustible) and on its neutronic properties as determined by the monitor. As an illustration, for the waste classification "combustible, low absorption, low moderation" (for example a 10 litre can of rubber gloves) the systematic error factor applied is 0.24; while for waste classification "non-combustible, high absorption, low moderation" (for example a 560 litre stainless steel drum containing ~750kg iron) the factor applied is 5. The limits of detection achieved in these examples are 88mg Pu-239 equivalent and 670mg Pu-239 equivalent respectively, values that are fully acceptable in this application.

Prior to each measurement the FMD carries out automatically a series of checks to confirm that its performance is acceptable. Included in these checks is confirmation of detector efficiency. This is carried out by measurement of a known neutron source which is moved into the field of view of the detectors by a neutron source transfer system. The source movement system has been produced by an external manufacturer to a Sellafield Technical Department specification. Similarly comprehensive checks are carried out, at the end of the measurement cycle, on the data generated to confirm that the measurement was carried out successfully.

Following the successful commissioning of the FMD, the next application of the DDA technique on plant will be in the THORP Hulls Monitor to determine the fissile content of waste arising from fuel shearing and dissolution.

4   CURRENT AND FUTURE DEVELOPMENT WORK

4.1 VLLW/LLW Sentencing Monitor

The monitoring of decommissioning waste at or around the VLLW/LLW boundary is a relatively new area of work for PS & ED group. To date this has involved studies into both the legislative requirements of uncontrolled disposal and also the feasibility of monitoring large volumes of waste with sufficient sensitivity to satisfy the present reuse/'free-release' requirements (taken as 0.4Bq/g over and above the natural activity of the waste material (6)). The monitoring systems developed will be required to satisfy a number of roles:-

(i) In-situ identification of material with activity below the VLLW/LLW boundary or that could be brought below this boundary by decontamination.

(ii) Sentencing of waste arisings into VLLW and LLW streams.

(iii) For waste identified as LLW, determination of the inventory information required for Drigg disposal.

For bulk concrete and brick waste, measurements are most likely to be based on the use of gamma spectrometry (both HRGS and LRGS) in a stand alone facility, taken in combination with inferential data obtained from both in-situ measurements and historical knowledge of the waste stream. Currently available neutron based techniques may be unsuitable for this particular application, due to the natural level of uranium in typical building rubble obscuring the signal from any fissile contaminants. However, this constraint will not necessarily apply in the monitoring of other waste streams, such as waste from steel structures.

The choice of in-situ monitoring technique will be dependent on the particular plant being decommissioned. Hence, the programme is aimed at developing all the necessary monitoring equipment to satisfy current and future decommissioning operations within BNFL. A number of radiometric techniques are being investigated for the purpose of in-situ monitoring, these include:-

HRGS - for direct sentencing of waste, under strict QA supervision, in situations where the monitoring geometry is favourable, such as a large flat wall.

LRGS/HRGS - to provide activity depth profiles in floors and walls by non-destructive means.

Beta-spectrometry - to enable the activity contribution of pure beta emitters to be inferred from activity ratio information, such as the Sr-90/Cs-137 ratio.

Passive X-ray spectrometry and X-ray fluorescence spectrometry - to provide supplementary information on gamma ray emitters with very low branching ratios, such as Pu isotopes.

The feasibility phase of this development programme is now largely complete, work is currently being directed towards overall system designs and the production of system specifications.

4.2 Other Development Areas

The examples given in Section 3 have satisfied or, in the case of the VLLW/LLW Sentencing Monitor above, are expected to satisfy customer requirements. However, new requirements for monitoring equipment to support specific monitoring needs are continuously being identified. Currently, work is in hand to develop systems for criticality and inventory measurements on alpha and beta-gamma waste items retrieved from temporary storage and arisings from decommissioning operations. As previously noted many measurement problems are common to different requirements and so, in addition to the work on specific measurement systems, a generic programme of work is being undertaken. Some of the problems requiring particular attention are outlined below.

(a) The measurement of plutonium in waste streams with a high neutron emission rate from alpha,n reactions. Work has been carried out in this area using HRGS.

(b) The derivation of methods of measuring U-235 in the presence of plutonium - this is a requirement of some current criticality clearance conditions.

(c) The extension of inferential methods for isotopic inventory determination to areas where the reactor bred isotopic correlations have been destroyed, ie. following reprocessing operations or storage under certain conditions.

(d) The development of methods to overcome current problems with thermal neutron interrogation techniques, eg. the DDA technique. These problems include:-

(i) The concentration effect whereby the method cannot cope with the measurement of concentrated masses of plutonium greater than a gram or so due to source self shielding unless the fissile material is effectively 'diluted' by non-fissile species such as U-238 in the fuel residues.

(ii) The effect of the matrix material and the position of the fissile material in the matrix on interrogation efficiency and detection sensitivity. A considerable amount of work has been carried out in this area in support of the FMD development, but more needs to be done.

(iii) The determination of ways of overcoming the variability of the active background in the DDA technique. This is an obstacle to accurate measurement of miscellaneous waste where the sample matrix differs from the calibration matrix.

5 CONCLUSIONS

Radiometric physics development has played a crucial role in the provision of instrumentation to satisfy the requirements of radioactive waste management at Sellafield. The Swarf Inventory Monitor, for determining the carry-over of uranium fuel and the fission product inventory of swarf arising from the decanning of Magnox fuel elements, has proved very successful and has now been formally handed over to plant. Similarly, initial operation of the Fissile Material Detector, for providing fissile material assays on miscellaneous beta/gamma waste packages consigned to the MBGWS, has been very encouraging. However, formal handover to plant is awaiting a change-over from the currently installed, laboratory standard, neutron generator to a newly developed industrial standard neutron generator.

In anticipation of future customer requirements, Physical Science and Engineering Development Group at Sellafield are currently involved in a number of development programmes, such as the development of monitoring systems for essentially 'non-active' waste. These programmes are aimed at refining and improving currently available monitoring techniques, to extend the range of available methods and to provide further measurement systems, via the methodology described, to meet specific requirements.

ACKNOWLEDGEMENTS

The authors would like to thank those members of PS & ED group who provided assistance in preparing this manuscript.

REFERENCES

(1) KNOLL, G.F., Radiation detection and measurement. (John Wiley, New York, 1979).

(2) SHER, R., and Untermyer-II S. The detection of fissionable materials by non-destructive means. (Amer. Nucl. Soc., Illinois, USA, 1980)

(3) CALDWELL, J.T., and KUNZ, W.E. Experimental evaluation of the differential die-away pulsed neutron technique for the fissile assay of hot irradiated fuel waste. ANS Topical Meeting on the Treatment of Radioactive Wastes, Richland, Washington, 19-22 April 1982.

(4) BURSTALL, R.F. FISPIN - A computer code for nuclide inventory calculations. AEA Report ND-B-328(R), October 1979.

(5) FOX, G.H., McDONALD, B.J., and GARDNER, N. The development of radiometric instrumentation in support of Sellafield plants. Int. Conf. on Nuclear Fuel Reprocessing and Waste Management - RECOD '87, Paris, 23-27 August 1987.

(6) 'Radioactive Substances Act 1960', HMSO, London.

Quality Assurance of waste packages to be disposed of at the Nirex deep repository

A R DAVIES, MSc and M J S SMITH, MA
UK Nirex Ltd, Harwell, Didcot, Oxfordshire
P BARTON, BEng, CEng, MIMechE and M W WAKERLEY, PhD, CChem, MRSC
ANS Consultants Limited, Epsom, Surrey

SYNOPSIS

UK Nirex Ltd. has been formed to develop and implement methods for the safe disposal of low- and intermediate-level radioactive waste. It is Nirex policy, as well as a regulatory requirement, that Quality Assurance be applied to all activities related to waste disposal.

This paper describes the Nirex approach to:
- Quality Assurance as applied to the packaging of radioactive wastes for disposal;
- Compliance Assurance;
- the establishment of a database that will contain information on the nature and contents of each waste package;
- Threshold Recording Levels for radionuclide inventories in the database;
- a system for uniquely identifying each waste package.

1 INTRODUCTION

UK Nirex Ltd. (Nirex) has been formed by the nuclear industry to develop and implement methods for the safe disposal of low- and intermediate-level radioactive waste (LLW and ILW). It is Nirex policy and a regulatory requirement that quality assurance be applied to all activities related to the disposal of the waste. This paper will describe the approach being taken by Nirex to quality assurance of the packaging of radioactive wastes for disposal, and the system of compliance assurance that the company has established. In addition the paper will describe the approach being taken to establish a database that will contain information on the nature and contents of each package of waste for disposal. The information to be recorded will include a radionuclide inventory, and to this end Nirex has developed a Threshold Recording Level (TRL) concept. The TRL defines the level at which each radionuclide might be deemed insignificant for regulatory, operational and safety reasons. Examples of the application of the TRL concept and details of the Nirex system for uniquely identifying each waste package will be given.

2 REGULATORY REQUIREMENTS

Her Majesty's Nuclear Installations Inspectorate (HMNII), a division of the Health and Safety Executive, has set licence conditions for the waste packaging facilities now operating. HMNII will be responsible for licensing operations at future nuclear facilities including the Deep Repository being developed by Nirex. HMNII has issued a guidance document (1) on quality assurance which gives their position on the detailed provisions that should be covered by a licensee's quality assurance arrangements, and on the methods of assessment and monitoring available to HMNII.

In addition to a site operating licence, an authorisation is required to dispose of radioactive waste. This is granted jointly by the Department of the Environment and the Ministry of Agriculture, Fisheries and Food, or their counterparts in the Scottish Office. The views of these authorising departments with regard to quality assurance of the processing of raw radioactive waste into an immobilised form which is suitable for land disposal, were first outlined in a 1984 information note (2). At that time their requirements for quality assurance relating to the disposal of waste were to be published in a separate document. Before granting an authorisation for the Deep Repository the authorising departments will require quality assurance arrangements to have been implemented.

The Department of Transport (Radioactive Materials Transport Division) acting on behalf of the Secretary of State for Transport, is the Competent Authority for all matters relating to the transport of radioactive materials. It administers internationally agreed regulations (3) which include specific requirements that quality assurance programmes shall be established for the design, manufacture, testing, documentation, use, maintenance and inspection of all packages. The Department of Transport periodically examines the quality assurance records of transport organisations, the performance of auditors and any other self-checking mechanisms related to the quality assurance programme, so as to judge the acceptability of the arrangements and hence provide evidence of regulatory compliance (4).

3 NIREX POLICY ON QUALITY ASSURANCE

It is the policy of Nirex that:

(a) For all items and systems which make up their waste management technical services for LLW and ILW, there shall be in force appropriate arrangements for providing assurance of quality at all stages from design to the monitoring of completed disposals.

(b) The arrangements made for assuring the quality of items and systems at the repository shall, wherever appropriate, satisfy the principles expressed in BS 5882 'Specification for a total quality assurance programme for nuclear installations'.

(c) Waste producers shall be required to satisfy Nirex, in such manner as the Company may determine, in regard to the nature and the proper performance of any treatment and packaging operations carried out by the waste producers which could affect the safe condition of the wastes during and after their transfer to Nirex.

This policy is implemented within Nirex by a quality system which is structured to comply with the requirements of BS 5882. In particular, a company quality assurance manual has been prepared which provides an overall description of the company's quality management system. This is supported by a series of company procedures. Quality assurance programmes for authorised projects or facilities will draw on the company quality assurance manual and the company procedures as relevant to the nature of the project or facility. If necessary, such programmes will be supported by arrangements

and procedures for specific projects or facilities, as is already the case for the transport of wastes. Where relevant, requirements for quality assurance, including compliance with the appropriate aspects of BS 5750 'Quality Systems', are specified for contractors working on behalf of Nirex.

4 NIREX QUALITY ASSURANCE REQUIREMENTS FOR WASTE PACKAGES

4.1 Main requirements

Waste acceptance criteria will be specified by Nirex once the repository has been granted an authorisation by the regulatory authorities. As an interim measure Nirex, in collaboration with the waste packagers, is preparing guidelines for the packaging of wastes and technical specifications for the packages themselves, in order that waste packagers can begin packaging their wastes.

The Nirex waste package specifications provide the essential link between the design of the waste packages and the design and safety case for the repository and associated transport system. Table 1 shows the proposed contents of the Nirex Waste Package Specification document which covers four standard LLW and four standard ILW containers (details of the specification of the 500 litre drum, the main container

Table 1 Proposed contents of the Nirex Waste Package Specification document

---

Chapter 1 Introduction

Chapter 2 Quality Assurance requirements

Chapter 3 Waste container specifications

Chapter 4 Data requirements

Chapter 5 Other specifications

Chapter 6 Ancillary information

Chapter 7 Explanatory material and design guidelines
- waste container specifications

Chapter 8 Explanatory material and design guidelines
- other specifications

Chapter 9 Guidance documents - miscellaneous
9.1 Quality Assurance

---

for ILW, are given in (5)). Chapter 2 of the document, which lays down quality assurance requirements, has been issued to the waste packagers and is being implemented. The main sections of this Chapter are now examined in more detail.

Requirements

Waste packagers are required to establish quality systems in accordance with the applicable elements of BS 5882. The appropriate criteria from this British Standard are to apply to all activities that have a bearing on the quality of the packaged waste. These activities cover process development, container design, waste characterisation, plant testing, commissioning, operations and aspects of plant design to the extent that they affect product quality.

Quality assurance programmes describing the systems are to be in place for activities affecting product quality. The programmes are to clearly identify those activities for which quality plans will be used and which procedures will be appropriate.

Quality plans describing the step-by-step sequence of quality-related activities are required. These are to describe in detail how a waste packager ensures and records that the technical requirements have been met in the course of the activity.

A waste product specification describing the quality (properties and composition) and performance characteristics is required for each waste package type. Where a packaging plant is intended to be used for different feed streams and production of different packaged waste streams, then sufficient details need to be provided so that there are no ambiguities.

A system, independent of the production personnel, of verification of a sample of waste package quality records is to be in place. This is to ensure that the packager has verified the adequacy of the quality of his product and that the quality records for packages are consistent with the product specification and approved quality plans.

Submissions

Quality assurance programmes, quality plans and waste product specifications are to be made available by the waste packagers to Nirex for acceptance. In particular, all necessary records are to be generated, maintained and made available to demonstrate that the appropriate requirements have been complied with, and all necessary data will be available for transmission to Nirex prior to transport of a package to the repository. Additionally, the quality plans are to ensure that appropriate levels of control have been applied in verifying conformance to requirements.

Assessments

Waste packagers are responsible for demonstrating that the necessary quality assurance arrangements are in place, that these arrangements are implemented and that any necessary corrective actions are taken. Nirex undertakes assessments and audits of the arrangements (in addition to those carried out by waste packagers) and has appointed Lloyd's Register Quality Assurance Ltd. (LRQA) as an independent external auditor to carry out these tasks on its behalf. It is intended that this use of an independent auditor will minimise multiple external auditing of packaging plants.

Waste packagers may alternatively choose certification of their quality systems. This will be acceptable provided that the certification requirements are equivalent to those of BS 5882, and also provided that the certification body is accredited by the National Accreditation Council for Certification Bodies and its scope is extended to include radioactive waste management. Nirex will require copies of relevant documents produced by the certification body and retains the right to carry out additional assessments should the need arise.

4.2 Guidance

Chapter 9 of the Nirex Waste Package Specification document will provide guidance and further explanation on quality assurance as applied to waste packaging systems. A typical contents list of a quality assurance programme for a radioactive waste packaging plant is presented in Table 2.

Table 2 Typical contents of a quality
assurance programme for a waste
packaging plant

---

1. INTRODUCTION

2. DESCRIPTION OF PACKAGING PROCESS
   - flow diagram showing process and QA hold
     points and covering radwaste feedstock,
     drums, encapsulant and finished product

3. RESPONSIBILITIES AND ORGANISATION
   - Authority and responsibility
   - Communications
   - Organisational interfaces
   - Commissioning
   - Staffing and training
   - Execution of QA programme
   - Plant safety and security
   - Staff safety
   - Maintenance

4. DESIGN CONTROL
   - Interfaces
   - Verification
   - Modifications

5. PROCEDURES, INSTRUCTIONS AND QUALITY PLANS

6. DOCUMENT CONTROL
   - Preparation, review and approval
   - Release and distribution
   - Change control

7. CONTROL OF PROCUREMENT, PURCHASED ITEMS
   AND SERVICES
   - Procurement documentation control
   - Identification of items (drums, etc.)

8. PROCESS CONTROL
   - Commissioning
   - Operations
   - Modifications

9. INSPECTION TESTING AND SURVEILLANCE
   - Receipt inspection
   - In-service inspection
   - Hold points
   - Test programme and performance
   - Calibration

10. NON-CONFORMING ITEMS
    - Radwaste feedstock
    - Drums
    - Encapsulant feedstock
    - Encapsulated material

11. CORRECTIVE ACTION

12. RECORDS
    - Preparation
    - Collection
    - Storage and maintenance

13. REVIEWS AND AUDITS
    - Internal
    - External
    - Management review

---

5 MEETING REQUIREMENTS ON QUALITY

5.1 General

In many industries the product from a
process is subject to detailed inspection
and testing both non-destructively and to
a lesser extent destructively. However,
detailed examination on a routine basis is
not technically feasible with immobilised
waste because of its radioactive nature and
because such tests might adversely affect
the suitability of the waste package for
disposal.

In order to ensure that all packages are
of the specified quality, the waste packagers
have implemented a number of measures.
Firstly, extensive research and development
has been undertaken to characterise the
envelope of process parameters which produce
packages meeting the requirements. Secondly,
quality systems have been introduced to
cover all activities having a bearing on the
quality of the packaged waste, and to ensure
the waste is packaged within the prescribed
parameters. These quality systems include
inspecting and testing the non-radioactive
components of the package i.e. the waste
containers and the encapsulant feed materials,
which in most cases represent the bulk of
the mass of the final product. Additionally,
as there is no technique capable of
accurately measuring the radionuclide content
of most wastes after packaging, detailed
inventory measurements are made on samples
of the waste (or the bulk waste itself)
prior to packaging.

Some other properties of the waste
package, while important in their own
right, will also be used to provide
information for quality assurance purposes,
e.g.

- identifier number, which with other
  records will identify the location,
  time and plant parameters during waste
  packaging;

- weight, which with other records will
  confirm correct filling;

- surface contamination level, which
  will constitute one check for misfilling
  or container damage in the unlikely
  event that such damage has occurred;

- external radiation level, which should
  be consistent with an expected value
  and enable the presence of certain
  gamma-emitting radionuclides to be
  verified.

All of these parameters can be determined
and recorded at different stages throughout

the packaging, transportation, and disposal sequence.

The regulatory authorities may wish to undertake non-destructive testing of some or all packages as additional compliance assurance. If destructive testing is undertaken then it can be on only a small percentage of packages, since any tested package would have to be brought back to within the waste package specification in order for it to be disposed of by Nirex.

5.2 Waste package database

The authorising departments have stated (6) that there must be adequate provision for detailed records of the wastes and their locations within the repository, and that these records are to be maintained and preserved in an agreed format. Nirex is defining what information on each individual package will have to be recorded. Figure 1 illustrates the material inputs and processes in package preparation and disposal.

Most of the data on a package will be generated by the waste packager at the packaging stage. This will include data on the nature and contents of the package, a radionuclide inventory, and certification of compliance assurance. The record will also include a history of the package and relevant information on storage and transport. Nirex will make some measurements upon receipt of the package and these will be used to verify compliance with the information

provided by the waste packager. Finally the location at which the package is emplaced within the repository will be recorded.

Nirex intends to set up a computerised database (the waste package database) to hold all the relevant information. This will be used to meet operational requirements as well as to provide a long-term record. It will be the definitive source of data on each package and will allow an accurate inventory of all waste disposed of at the repository to be calculated. The information to be recorded is that identified as being necessary to meet a defined need. Most of the data will need to be stored beyond the time of repository closure whilst some, such as surface contamination levels upon shipment and repository operational information, might only need to be kept until vault closure.

There should be no need to duplicate the detailed records of the operation of packaging plants and stores which will be held by waste packagers. Once compliance of a package with the product specification has been established, these waste package records will be of relevance to Nirex only in exceptional circumstances. The data to be held post-closure by Nirex will also be recorded in traditional archival form. Records will be maintained in a manner that allows ready retrieval and prevents loss of any particular record, and they will be kept in an environment that minimises deterioration and damage.

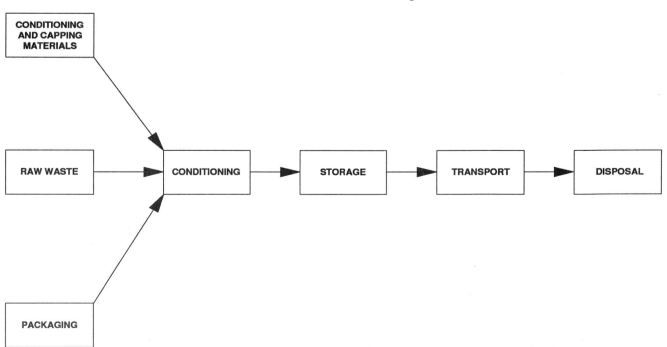

Fig 1   Material inputs and process in package preparation and disposal

6 THRESHOLD RECORDING LEVELS

In order to ensure that the radionuclide inventory for each package contains data for all the radionuclides that are present at levels of significance, whilst avoiding unnecessary data requirements for radionuclides that are below the levels of significance, Nirex is in the process of defining for each radionuclide and each package type a 'Threshold Recording Level' (TRL). The waste packager will have three choices when recording the quantity of a particular radionuclide or the total alpha or total beta-gamma activity in a particular package. These will be:

- not present (i.e. cannot possibly be present due to the composition or nature of the materials handled);

- less than the TRL;

- a numerical value, obtained to the required accuracy by measurement or inference.

The TRLs for the concentration of each radionuclide in a 500 litre drum of ILW have been defined by the following procedure:

(a) Identifying in broad terms the situations in which the radionuclide inventory of the drum influences the safety of the waste transport, handling and disposal system, or is otherwise of significance.

(b) Establishing the safety assessment criteria, operating limits or other factors appropriate to each situation.

(c) For each situation, deriving estimates of the quantities or concentrations of individual radionuclides which, if present alone, would become a limiting quantity.

(d) For each situation, assessing and then applying the factor by which this limiting quantity would have to be reduced in order to produce a concentration considered insignificant. Nirex have set design targets for doses and risks during transport, handling and disposal that are well within regulatory levels, and thus levels of radionuclides which can be considered insignificant to Nirex are also likely to be low in absolute terms.

(e) Defining the TRL for each radionuclide of interest as the lowest of these 'insignificant' concentrations.

A similar approach is being taken to defining the accuracy to which the activity value for each radionuclide is required. This inevitably depends on its concentration in relation to the TRL.

For the TRLs eight situations have been considered and their relevant safety criteria and limiting quantities defined (stages a, b and c above). This procedure is illustrated for three of these situations relating to the long-term safety of waste disposal.

Preliminary assessments have been undertaken by Nirex of the long-term risks due to human intrusion, groundwater pathways and gaseous releases for a deep repository sited at Sellafield or Dounreay. From these assessments, for each of the three modes of exposure, limiting quantities and concentrations have been derived for each relevant radionuclide which, if present alone, would just reach the appropriate DoE target criterion of a risk to an individual in a year equivalent to that associated with a dose of 0.1 mSv. For about half the radionuclides studied their groundwater concentrations, and hence the associated risks, are limited by their solubility under repository conditions. For most of the radionuclides the risk is equivalent to less than one thousandth of the DoE target criterion so the long-term safety of disposal of waste imposes no limit on the concentration of these radionuclides in the waste.

The other five situations considered related to transport, heat generation, nuclear criticality safety, normal package handling and an accident during package handling. Having arrived at limiting concentrations for all the situations for each relevant radionuclide, a composite factor was applied to take account of the number of radionuclides likely to contribute to a particular situation, and also to take account of the uncertainties in the derivation of the limiting concentrations.

In radiological assessments it is considered rigorous to use as complete a radionuclide inventory as possible, and to use upper-limit estimates of the minor radionuclides for which reliable estimates are not available. Thus if the level at which a radionuclide is considered insignificant were to be set as high as 0.01 of its limiting concentration, then an assessment which includes 30 unquantified radionuclides would yield a minimum impact of 30% of the limit, before allowing for uncertainties and before the quantified contributions from the major radionuclides were included. Such arguments lead to factors of insignificance being provisionally set at 0.001 to 0.01 of the limiting concentrations for different situations, thus allowing calculations of the insignificant concentrations and completion of stage (d). It may be possible to modify these values as the safety cases are developed.

Stage (e) was straightforward because the TRL was identified for each radionuclide as the lowest of the insignificant concentrations. Since the TRLs do not need to be cited to a great degree of accuracy it was convenient to round to the nearest order of magnitude.

The preliminary results of the calculations for nearly 70 radionuclides revealed that a gaseous pathway, which may occur after disposal, was the most limiting for H-3 and C-14, whereas the groundwater pathway after disposal set the TRLs for eight other radionuclides. The transport situation sets TRLs for over half the radionuclides, whilst the intrusion and heat generation situations did not set any TRLs and the criticality situation set only one TRL. The shielding situation set TRLs for about a quarter of the radionuclides, as might be expected since these possess high gamma energies, typically >0.6 MeV. The handling accident situation set TRLs for two of the lowest specific activity radionuclides Th-232 and U-238.

A comparison is now being undertaken between the values of the derived TRLs and the concentrations of the radionuclides in the various waste streams for which detailed data is given in the National Inventory (7) in order to identify improvements that can be made in the methodology. Further work is planned by Nirex to refine the TRL concept and to define TRLs for all the standard waste packages.

7 WASTE PACKAGE IDENTIFIER

An essential element of the waste package database is the waste package identifier which is marked on each waste container. The identifier has been designed to provide unique identification for each waste package and forms the vital link between any waste package and the database. The identifier permits the waste package to be correctly recognised and recorded during all stages of conditioning, storage, handling, transport and final disposal.

The waste package identifier comprises ten alphanumeric characters in three data fields. The first two characters are hexadecimal and are used to define the waste producer. Use of hexadecimal numbers provides 256 different combinations which are sufficient to permit major sites of arising to be identified separately. The following six characters are numeric and form a package number from 000001 to 999999. The last two characters are numeric and are check-digits designed to permit automatic checking that the number entered has been read correctly and is a valid number. The check-digit system has itself been subjected to detailed assessment to ensure that it can reliably detect common reading errors such as transcriptions and various types of transposition.

The OCR-A (Optical Character Recognition type A) character has been specified as the format for marking identifiers onto waste packages. The advantage of this system over others e.g. bar-codes, is that it permits both man and machine reading of the identifier. Various techniques are available for attaching identifiers to waste packages, such as laser etching or labels, but all identifiers must meet stringent requirements for longevity and ease of recognition.

8 CONCLUSIONS

It is the Nirex policy that quality assurance be applied to all its activities relating to waste disposal. The company has specified quality assurance requirements for the packaging of radioactive wastes for disposal and has established a system of compliance assurance. Nirex will hold information on the nature, contents and location of each package of waste disposed of in the Deep Repository and is in the process of specifying the information to be held on a waste package database. This information will include an inventory of the radionuclides in each package; a methodology is being developed for defining the concentrations (TRLs) below which the presence of each radionuclide would be deemed insignificant and therefore would not have to be recorded on the database. Nirex has also established the system for uniquely identifying each waste package.

REFERENCES

(1) Health & Safety Executive, Nuclear Installations Inspectorate. A guide to Quality Assurance Programmes for Nuclear Installations: NII/R/8/85, March 1985.

(2) Department of the Environment. Radioactive Waste Management: Information Note No.2, 1984 – Quality Assurance in Processing Radioactive Wastes for Land Disposal.

(3) International Atomic Energy Agency. Regulations for the Safe Transport of Radioactive Material (1985 edition as amended 1990), IAEA, Vienna 1990, Safety Series 6.

(4) Pecover, C.J. Quality assurance in radioactive material transport. Nucl. Energy, 1990, 29, No.5, Oct., 367-373.

(5) Smith, M.J.S., Sievwright, R.W.T., Holt, G. and Donelan, P. The Packaging of Waste for Disposal in a Deep Repository. Radioactive Waste Management 2, BNES, London, 1989.

(6) Department of the Environment, Scottish Office, Welsh Office, Department of the Environment for Northern Ireland, Ministry of Agriculture, Fisheries & Food. Disposal facilities on land for low and intermediate level radioactive wastes: Principles for the protection of the human environment. HMSO, 1984.

(7) Electrowatt Engineering Services (UK) Ltd., The radionuclide content of radioactive waste, UK Nirex Report No. 95, December 1990.

C431/052

A computerised data acquisition system for Quality Assurance in an intermediate level radioactive waste encapsulation plant

A J KEEL and J D HILL, MA
AEA Technology, Winfrith, Dorset

SYNOPSIS

The Winfrith Radwaste Treatment Complex will process intermediate-level waste in both solid and sludge forms. Waste will be encapsulated within a cement matrix and sealed in 500-litre stainless steel drums in the Radwaste Treatment Plant (RTP); it will then be transferred to the Treated Radwaste Store (TRS) for storage, to await final disposal in a UK NIREX Ltd (NIREX) repository.

A database system is required to provide a complete Quality Assurance record for encapsulated waste, in order to satisfy NIREX requirements. This is implemented using the ORACLE relational database management system, running on a MicroVAX 3100 computer.

INTRODUCTION

The Radwaste Treatment Complex is operated by AEA Decommissioning and Radwaste, one of the businesses of AEA Technology (the trading name of the United Kingdom Atomic Energy Authority), and is situated at the Winfrith Technology Centre in Dorset. The Winfrith site has 30 years experience of nuclear power generation, research, development and technical services in both nuclear and non-nuclear fields. The Radwaste Treatment Plant (RTP) is designed to encapsulate intermediate level wastes (ILW) arising from nuclear activities.

The waste is immobilised in a form suitable for disposal in the NIREX repository. Assurance is required that each drum of encapsulated waste meets the repository acceptance criteria. This paper explains the computer system which is being developed to gather data to provide this assurance.

1. BACKGROUND TO SYSTEM

1.1 The Waste

The Radwaste Treatment Complex will encapsulate ILW in both solid and sludge forms.

Sludge wastes (aqueous slurry comprising 25-30% solid content) have accumulated from operation of the Winfrith Steam Generating Heavy Water Reactor (SGHWR). These sludges are generated by decontamination of primary cooling circuits

and water treatment processes. The sludges are stored in three tanks, and are characterised by chemical and radiochemical analysis.

Solid wastes arise from the SGHWR building and from Active Handling operations (post irradiated fuel examination); solid ILW typically consists of tools, clothing, cans, glass, swabs and swarf. Solids are stored in sealed mild steel cans, which are held in shielded containers. The source of each can and its surface dose rate are known; this information is currently stored on a computer elsewhere on site. Detailed information about the contents may only be obtained by opening the cans.

1.2 Functions of the Complex

The Radwaste Treatment Complex consists of the following facilities:
- the Radwaste Treatment Plant (RTP)
- the Treated Radwaste Store (TRS)
- the Radwaste Operations Building (ROB)

The RTP has been developed to encapsulate solid and sludge waste streams. The waste will be immobilised within a cement matrix in 500-litre stainless steel drums, in a form suitable for final disposal in the NIREX repository.

The TRS has been developed for interim storage of the drums prior to eventual transfer to the national repository. This repository is to accept low and intermediate level waste, and is to be situated deep underground. It will be available within 25 years. The ROB is for remote management of the RTP and TRS, including control and storage of data. The computer is located in this building, and communicates with the RTP and TRS using terminal servers over the site Ethernet.

1.3 Processing of waste

Solid waste arrives at the plant in sealed mild steel cans within shielded overpacks; empty NIREX drums (each with a unique hexadecimal NIREX identifier) and cement powders are also delivered. The following processing occurs:

- the steel cans are removed from the overpacks
- the waste is sorted and placed in a NIREX drum; this is performed in a Sort Cell using master-slave manipulators
- a mix of grout is made from cement, aggregate and water
- the drum is filled with grout
- the mix is allowed to cure
- the lid is welded on
- the drum is smeared to check for any contamination
- the completed drum is placed in an overpack and taken to the store.

The procedure is similar for sludge waste, except as follows:

- sludge is delivered to the plant in NIREX drums, ready for encapsulation
- dry powder is added instead of grout, and is mixed in the drum

Each stream of waste processed by the plant, be it solid or sludge, will define a separate campaign. Each campaign will be divided into one or more sub-campaigns, according to the cement recipe used; this allows for changes in powder batches. Cement and aggregates are stored in silos adjacent to the RTP.

1.4 Control of processing

The RTP is controlled by a General Electric Series Six Programmable Logic Controller (PLC). This contains programmed sequences for plant operation, controlling movement of drums and waste; software interlocks prevent invalid operations. Powder movements are controlled by a separate grout plant PLC, which ensures consistent ratios of ingredients. A Unimation 762 robot welds the lids onto the drums using a MIG welder and performs health-physics related operations (smearing, measurement of contamination etc). A British Federal ARCGUARD Weld Monitor is used to provide data on the welds. The robot

sequence is initiated by the PLC.

1.5 Data

The processes described above generate large quantities of data, which will be required to assure NIREX of the quality of the completed drums before final disposal. A computer database is therefore required, fulfilling the following functions:

- Provision of data to satisfy NIREX on the quality of the product without post-production testing. This will include information about the waste itself (origin, type, composition, activity), materials used for encapsulation (drums, cement, powders), evidence of processing (measurements, events and movements) and drum storage information
- Provision of data to assist local management in achieving a quality product, including management reports and a comprehensive Planned Maintenance system

2. SYSTEM REQUIREMENTS

2.1 General requirements

The major function of the system is to gather data transmitted from plant instrumentation, to store the data and to present it to users. The data is to be stored in both its raw (as transmitted) and converted (engineering units) forms. The system must cross-check independent measurements of parameters where available, notifying the plant operators of any inconsistent or out-of-range values.

A number of values are not supplied directly by the instrumentation, and will be obtained from users at terminals. Such inputs must be verified and validated wherever possible.

All plant control functions are exercised from the PLC; the role of the computer is restricted to data gathering.

All data relevant to the quality of encapsulated waste must be kept until two years after final disposal in the NIREX repository. This implies a data life of up to 30 years on current estimates. The detailed requirements can be expected to change; any system implemented must have the flexibility to cope with this.

The integrity of all stored data must be guaranteed; this includes resilience against tampering and against loss of storage media. Data gathering must be continuous, and must not be held up to wait for user responses. Data transmitted from the plant must not be missed in the event of system or software failure; data gathering must restart automatically as soon as possible.

2.2 Data requirements

2.2.1 Drum Data

The system must store a full record of the processing undergone by each NIREX drum, including the time at which significant actions occurred. All quality-related parameters must be kept. The waste content of the drum must be recorded, including radionuclide inventory and supplier. For solids, this involves material, mass and volume of items. For sludge, chemical and radiological analyses are required, together with the identity of the analyst.

2.2.2 Campaign Data

The computer must record initiation and completion of campaigns and sub-campaigns, storing characteristics of the waste streams and cement mix recipes.

2.2.3 Materials Data

The system maintains a record of delivery and usage of consumables used in the production of completed NIREX drums. These items include the empty drums, cement and aggregate powders, weld wire and argon. This information is kept as part of the records for each drum, and also for management.

2.2.4 Instrument Data

The quality of information about drum processing depends heavily on the reliability of data from the plant instrumentation. Periodic calibration on all instruments must be performed and recorded; derivation of each drum parameter from more than one source, where available, will assist with reliability checking.

2.2.5 Treated Radwaste Store data

Surface dose rates are measured by the Bulk Gamma Monitor for every drum and transmitted to the computer. A complete movement history is maintained for each drum within the store. Periodic smear checks will be performed on drums and monitoring results recorded.

2.2.6 Maintenance Data

Plant management and maintenance are to be assisted by the computer system. The requirements include:
- Plant components and asset listings
- Plant performance and failure analyses
- Spares inventory
- Drawing register
- Staff deployment and scheduling
- Maintenance schedules, history and instructions
- Financial data, resource usage analysis

3. SYSTEM IMPLEMENTATION

3.1 Hardware

The system runs on a Digital Equipment Corporation (DEC) MicroVAX 3100 Model 20e running the VAX/VMS operating system. This offers the following important features:
- a real-time programming capability
- a full multi-tasking capability
- six-hour callout on faults to minimise down-time
- high level of future support and maintenance commitment (DEC undertake to maintain all equipment for 10 years after withdrawal)
- an upgrade path
- local experience of the equipment
- Ethernet connectivity to remote locations

Given the long lifetime required for the data, major changes may be required; selecting hardware with a long lifetime will delay and minimise disruption.

3.2 Software

The software to record information is based on the ORACLE Relational Database Management System. The real-time software to gather data is written in DEC Fortran-77. The volume and complexity of data to be generated by plant operations imply the need for a database product. The relational approach is well-established, and offers a pathway into the future. Relational technology allows data to be stored without prejudice on future queries. Use of the standard Structured Query Language (SQL) for database query and manipulation allows data interchange with other products where required.

Fortran is a well-established language for technical work and the VAX/VMS operating system offers suitable facilities for real-time programming.

The ORACLE product offers the following major advantages:
- robustness; the data cannot easily be corrupted
- a wide range of hardware platforms
- high vendor commitment
- an interface between the database and Fortran programs (via embedded SQL calls)
- ORACLE is the AEA Technology standard, giving data interchangeability with other local

systems
- Rapid fourth-generation program development facilities
- CASE (Computer Aided Software Engineering) tools are provided for software development

The major technical challenge of this project was insertion of real-time data from a working plant into a relational database.

3.3 Sources of data

3.3.1 The Plant PLC

The plant PLC is the main source of data. Data is transmitted to the MicroVAX by an ASCII-Forth Module (AFM; an interface board) within the PLC. The AFM contains a Forth program which communicates with a computer via a serial line using the industry-standard Modbus protocol. The computer can request values from registers held within the AFM; the registers can hold plant data or status flags, and always contain unsigned 16-bit numbers.

A number of "events", together with associated data, are defined within the AFM. When one of these occurs, a record of the event and its data is written to a circular buffer in the AFM; a pointer to the most recent event is maintained. When the value of the pointer changes, the data for a new event is available in the buffer. Events are designed so as to permit derivation of quantities with cross-checks where possible. Typical events are:

Event 9: Request delivery to drum at Grout Capping Station
Data: Drum Identifier (PLC-generated sequence number)
      Drum Weight
      Grout Temperature

Event 10: End of delivery to drum at Grout Capping Station
Data: Drum Identifier
      Pump Tank Weight
      Drum Weight

All data is supplied by the PLC in the form of integers in the range 0-4095; calibration data is stored in the computer.

Note that the computer may only request data from the AFM; no control function is provided.

3.3.2 Sort Cell Keyboards

Customised keyboards are provided at four locations for entry of data relating to the contents of solids cans. These are connected to the computer via a terminal server using RS232 serial lines. They are programmed to send ASCII characters and text strings to the computer, and to provide operator feedback using a liquid crystal display.

3.3.3 ARCGUARD Weld Monitor

The ARCGUARD monitors arc voltage, welding current, dip resistance and elapsed weld time. A weld is considered to have failed if these parameter values are outside programmed limits or if the percentage of arc instability is excessive. The ARCGUARD sends data over a serial line; this takes the form of an end-of-weld report, indicating pass/fail, parameters as above and a graphic "weld mimic".

3.3.4 Gamma scanner

A segmented gamma scanner may be used to obtain more detailed radiological information on the unopened solid waste cans and on individual items when manually sorted.

3.3.5 Character Recognition System

The RTP and TRS are equipped with character recognition systems to read NIREX identifiers on drums.

3.3.6 Users at terminals

Users at terminals (e.g. DEC VT220) enter some operational information (e.g. data on powders etc.) and all management information (e.g. asset, drawing and maintenance data). Where possible, input must be checked and verified.

3.3.7 Treated Radwaste Store

The computer will be required to interface to a PLC and a bulk gamma monitor (BGM) when the TRS becomes operational. The PLC will provide information on drum locations and movements within the TRS. The BGM will provide dose rates for each NIREX drum entering the TRS.

3.4 Software Architecture

The software is based on a central data repository in the ORACLE database on the MicroVAX computer; all of the information described above is stored here. Most of the data is gathered automatically and inserted into the database by a suite of real-time processes. Information is made available to system users via a system of screen menus and forms based on ORACLE technology; this interface is also used for some data input.

This architecture of three major components (database, real-time software and a user interface) is shown in figure 1.

3.4.1 The Database

The data model for the Radwaste Treatment Plant system consists of approximately 40 tables. Central to the system is the DRUM table; for each DRUM, we store its NIREX identifier, campaign and sub-campaign, date of processing and other items obtained from the plant PLC.

Processing data from the PLC is stored in two tables: EVENT and EVENT_DATA. Each EVENT is related to one drum, and can have one or more items of data associated (stored in raw form and as engineering units); from this, we can construct a history of the drum, complete with calculated parameters.

3.4.2 The Real-time Software

Acquisition and storage of data from the plant is achieved by a suite of real-time programs. These are controlled by the system manager, and run without further intervention as background processes. Supervision of the processes is achieved by a master process RT_SUPER; this provides resilience against software and communication failures by restarting any process which stops.

The processes gather data from the plant PLC, the Sort Cell keyboards and the ARCGUARD weld monitor. (Gamma Scanner and Treated Radwaste Store data and the Character Recognition System will be incorporated later.)

Gathering and storage of data is performed by separate processes in order to avoid delays in collection due to interaction with the database (which is comparatively slow). If errors are detected, the processes send messages to specified terminals, but data gathering is not interrupted.

A number of processes require data from the PLC. A task PLC_SERV performs all communication with the PLC; this accepts requests for data from other processes (via the VAX/VMS mailbox facility), communicates with the PLC and directs the reply to the requesting process.

See figure 2 for a diagram of the real-time software architecture. This software comprises the following processes:
- RT_SUPER, to supervise operation of the real-time suite
- PLC_SERV, to obtain PLC data for other processes
- FETCH_EVENT, to poll the PLC for events every 5 seconds
- DRUM_TRACK, to poll the PLC for drum movements and generate events
- FILE_EVENT, to write data from FETCH_EVENT and DRUM_TRACK to the database (using ORACLE Pro*FORTRAN statements)
- QA, to derive drum parameters from the stored event data (using ORACLE Pro*FORTRAN statements)
- ARC_COMM, to collect data from the ARCGUARD weld monitor
- ARC_DB, to write data from ARC_COMM to the database (using ORACLE Pro*FORTRAN statements)
- ALL_SORTS, to supervise operation of the sort-cell tasks
- SORT_1,SORT_2,SORT_3,SORT_4 to communicate with the four sort cell keyboards and write information to the database (using ORACLE Pro*FORTRAN statements)

The PLC stores the last 64 events and their data in a circular buffer, maintaining a pointer to the most recent event. The computer maintains a pointer to the last event successfully stored; this pointer is updated and stored as the event is written to the database. The FETCH_EVENT process monitors the PLC pointer, requesting event data whenever the two pointers are not equal, i.e. when new events have occurred. Thus, events are not lost if data gathering is disrupted; they are collected when normal service is resumed.

3.4.3 The User Interface

The user interface must allow inspection and input of data while protecting the database from accidental damage and unauthorised access.

Facilities are provided to the user through a hierarchic menu system, developed using the ORACLE SQL*Menu product. Each menu option gives access to a further menu or a screen form. Forms are developed using the ORACLE SQL*Forms product. Facilities are provided for verification and validation of data.

The top-level menu divides the system into a number of functional areas:
- Campaign Management, including campaign initiation, powders and consumables stock control
- Drum Management, displaying details of drums, their contents, movements and processing
- Radwaste Management, including solids/liquids stock control
- Engineering, including the Planned Maintenance system, drawing, asset and instrumentation registers
- System/Database Management, including control of real-time software and event definitions (only available to system managers)

Each of these areas also provides hard-copy reports using the ORACLE SQL*Plus product. Forms and menus all provide on-screen help as required; manuals give more comprehensive information. Users require brief training in keyboard use, but specialist computer knowledge is not required.

3.5 Software Development Strategy

3.5.1 Project Organisation

The plant is operated by Radwaste Operations Division of AEA Decommissioning and Radwaste, who are the customers for the data acquisition system and its interrogation facilities. Software development has been contracted to the Management Computing department of Winfrith Site Operations.

3.5.2 Method

The software is developed through several phases, from requirements specification through a general design specification to a series of functional specifications for each component of the system. Diagramming techniques (e.g. entity-relationship models) have been used as necessary.

Once a sound database design existed, much of the software was implemented using the ORACLE facilities, allowing rapid development. A user interface was developed through a series of prototypes, resulting in a system which met user requirements.

The newly enhanced ORACLE CASE tool will be used to aid future developments of the system. Existing software will be retrofitted to the CASE tool.

3.5.3 Software Quality Assurance

The software is being developed in accordance with a QA plan for the project and with departmental procedures. This involves production and independent verification of documents for each stage of the development, checking each document against the previous stage. Quality-related documents are controlled. Software is verified by independent inspection of code and validated by testing against a written test specification.

A change control procedure has been written to govern amendments to software, hardware and documentation. This procedure allows any system user to make observations and suggestions, while preventing unauthorised changes.

3.6 Security

Security is enforced through the username/password protection provided by the VAX/VMS operating system. A further level of security is provided by ORACLE username/password protection, controlling access to the database. Additional facilities are available to the database manager to give each ORACLE user access (read, input or update) to specified database tables and menu options.

Radwaste system users are split into classes according to the functions they require. These classes control access to specific tables and menu options.

Database backups (to a remote location) are performed automatically on a daily basis.

CURRENT STATE OF PROJECT

The ORACLE database system has provided a highly flexible platform on which to develop a data acquisition system, enabling storage of data without imposing any constraint on future enquiries. Capture of data from plant instrumentation has been demonstrated; the system is ready to provide evidence of the processing of drums. Development of registers for engineering drawings and plant assets have shown the power of the system as a plant management tool.

Recent inactive commissioning trials on the plant have tested components of the data acquisition system. Ongoing trials are being used to confirm the reliability of the overall system.

Figure 1: Radwaste Database System Architecture Summary

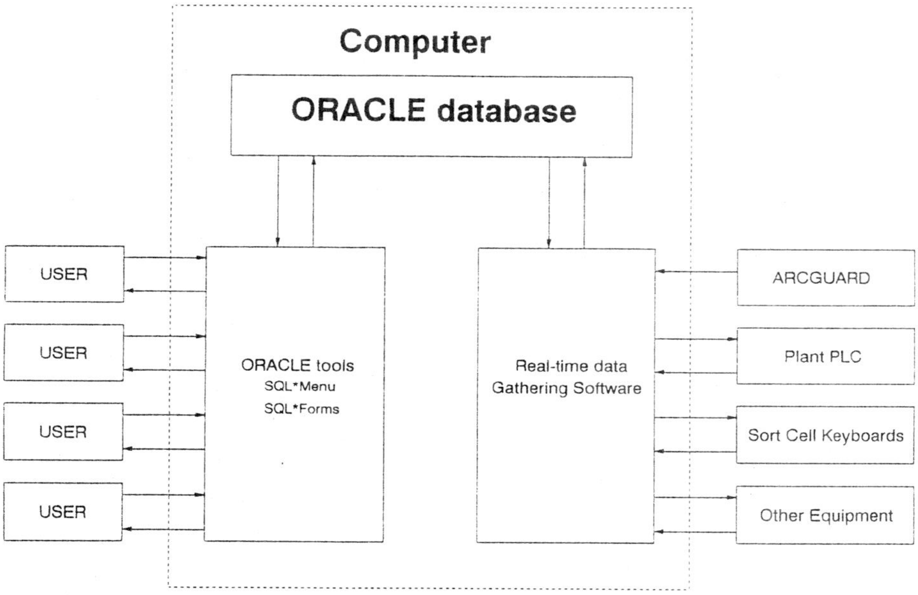

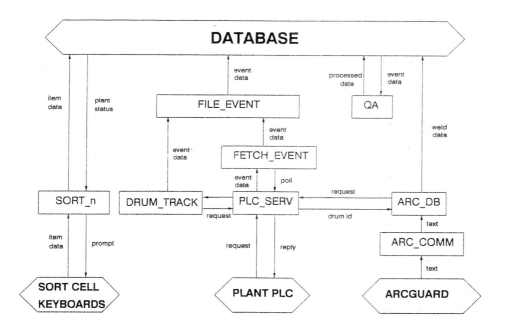

Figure 2: Radwaste Database Real-time Software Architecture

Regulatory framework for IP-2 freight containers

M C JANICKI, BSc, PhD and R A VAUGHAN, BSc, PhD
Croft Associates Limited, Upton, Didcot, Oxfordshire

SYNOPSIS

The publication by the IAEA of the 1985 Edition of Safety Series No 6 (IAEA SS6) Regulations for the Safe Transport of Radioactive Material [Ref 1] introduced performance requirements for packagings for a wide range of low activity bulk materials designated as Low Specific Activity (LSA) material or Surface Contaminated Objects (SCO). Materials which could be transported in simple packaging (such as drums or boxes) which had few regulatory requirements under the superceded 1973 Edition of IAEA SS6 [Ref 2], were now required to be carried within tested and approved packagings. After a review of options, it was clear that the most economic packagings for such materials were the largest packagings that could be used within transport weight and dimensional limitations. Freight containers of length of 20' and width of 8' and height of 4' and 8'6" were identified as the most appropriate.

Design specifications were developed for freight containers to meet the requirements of the IAEA Transport Regulations, and to maximise the technical and commercial benefits offered to consignors by this type of container for the transport and handling of bulk LSA/SCO materials.

This paper discusses the regulatory considerations which had to be reviewed and interpreted in developing freight containers as Industrial Package Type 2 (IP-2) [Ref 1] packagings, and the development of performance standards required to meet the regulatory requirements.

BACKGROUND

In 1987, British Nuclear Fuels plc (BNFL) as a result of a major programme of improvements in the method of disposal of LLW at the Drigg site in Cumbria , introduced the requirement for containerisation of all LLW to be consigned to Drigg for disposal. In the UK, Low Level Waste (LLW) is defined as waste having a radioactivity content of not more than 12 GBq/tonne beta/gamma and 4 GBq/tonne alpha. Such LLW does not normally require to be shielded during normal handling and transport.

Freight containers were developed for the transport of this LLW. It was recognised that such freight containers would have a wide application for radioactive material categorised as LSA material or SCO. The vast majority of freight containers in use are for non-radioactive materials transport. However, the benefits offered by the use of freight containers for the transport, handling and disposal of large volume consignments of radioactive material were recognised and subsequently adopted by the UK Nuclear Industry for the transport of LLW, and its disposal at Drigg.

The advantages perceived in using freight containers were that they could be manufactured economically and could be handled and transported using well developed systems. Freight containers are relatively low cost items, this being due primarily to the large numbers produced and the well established and standardised manufacturing methods employed.

In 1987 when the use of freight containers for the transport and disposal of LLW was introduced, either the 1973 Revised Edition (As Amended) or the 1985 Edition of the IAEA Transport Regulations could be used in the UK. Two designs of special freight container were developed to carry the different forms of radioactive material to be transported under the different Editions of the IAEA Transport Regulations then in use.

Under the 1973 Edition of the IAEA Transport Regulations, LLW was generally classified as Low Specific Activity (LSA) material which required packaging in an 'industrial package'. For this packaging standard the requirements were minimal and easily satisfied and did not include any testing or specified containment criteria. Some LLW of a higher activity level was classed as Low Level Solids (LLS), for which the packaging standard was a Strong Industrial Package. The requirements for a Strong Industrial Package included retention of contents following performance tests.

In order to transport packaged LLW, a 20' x 8' x 8'6" high freight container, with a conventional double door at one end, was designed to meet the requirements of the 1973 Edition of the IAEA Transport Regulations. This package was intended primarily for LSA and/or LLS materials (as drummed or wrapped items) which in combination with the freight container met the required performance standard of a Strong Industrial Package under the 1973 edition of the IAEA Transport Regulations.

Under the 1985 Edition of the IAEA Transport Regulations, LLW is generally classified as Low Specific Activity (LSA) material or Surface Contaminated Objects (SCO). LSA material by virtue of its nature has a limited specific activity; for LSA-II the average specific activity for solid material must not exceed 10^{-4} A_2/g, and for LSA-III solids it must not exceed 2×10^{-3} A_2/g. For SCO, which is a solid object which is not itself radioactive but has radioactive material distributed on its surfaces, limits for fixed and non-fixed contamination are defined in the Regulations. Under 'exclusive use' transport (which is always applicable for large freight containers), the packaging standard required is Industrial Package Type 2 (IP-2). The 1985 Edition of the IAEA Transport Regulations specify a performance standard in terms of package testing and containment criteria that is generally more onerous than that specified in the earlier 1973 Edition of the IAEA Transport Regulations for the same radioactive contents. The main requirements for an IP-2 package are that it shall protect against the loss or dispersal of the radioactive contents and loss of shielding under routine and normal conditions of transport.

A 20' x 8' x 4' high top opening freight container, as shown in Figure 1, was designed and manufactured to meet the requirements of an Industrial Package Type 2 (IP-2) as specified in the 1985 Edition of the IAEA Transport Regulations. This freight container was designed to carry loose radioactive solid material classed as LSA-II, or LSA-III or SCO.

In addition to meeting the IAEA regulatory requirements for transport both the 20' x 8' x 8'6" high and the 20' x 8' x 4' high freight containers were designed to meet the special needs for disposal at Drigg. Additional features were required to be introduced into the design of the freight containers: these included additional load bearing posts and base modifications, both of which were required to ensure that a stack of containers would transmit reduced loadings to the vault floor.

Since these two freight container designs were introduced, other variants of freight container have been designed and developed as IP-2 packages to meet the requirements of the 1985 Edition of the IAEA Transport Regulations. A further example, as shown in Figure 2, is a 20' x 8' x 8' 6" high version having a single sealed door at one end and designed as a reusable freight container for the transport of drummed material classified as LSA-II, or LSA-III.

The performance requirements for IP-2 freight containers, designed in accordance with the 1985 Edition of the IAEA Transport Regulations, are discussed in the following sections.

PERFORMANCE STANDARDS

In the 1985 Edition of the IAEA Transport Regulations, performance standards are specified for IP-2 packages which include design and test requirements. These requirements are intended to ensure that packages would not be significantly affected by conditions likely to be encountered in both routine transport, and normal conditions of transport including minor mishaps. For IP-2 packages, specific design requirements are stated (IAEA SS6 para 134):

- for a package (other than a tank or freight container)
- for a tank
- for a freight container

In the early stages of the development of the special freight containers, and prior to the introduction of the 1990 Amendment, there were interpretational difficulties with the 1985 Edition of the IAEA Transport Regulations. The main difficulty arose in the application of para 523 which is specific to freight containers. Para 523 stated, prior to the 1990 amendment, that freight containers may be used as IP-2 provided that they satisfy the requirements of ISO 1496/1-1978 [Ref 3] without loss of shielding integrity greater than 20%. However, there was no specific containment criteria specified within para 523 or the ISO tests; the water spray test specified in the ISO tests (1496/1-1978) being a rainwater in-leakage test not a containment test. The omission of a containment criteria was clearly at variance with the intent and philosophy of the regulations for this standard of packaging. The Explanatory Material of the 1985 Edition of the IAEA Transport Regulations [Ref 5] stated, prior to the 1990 Amendments, that release of contents considerations for IP-2 packages impose a containment function on the packaging for normal conditions of transport.

Due to the anomaly in the containment criteria of Paragraph 523, further unrelated concerns were raised as to the equivalence of the test conditions of the ISO tests (which are static tests) compared to the dynamic drop tests required for other industrial packages (excluding freight and tank containers).

This resulted in changes to the mechanical tests for freight containers being proposed, but these were not accepted in the IAEA review process which lead to the 1990 Amendment of the IAEA Transport Regulations. It was recognised that the static tests devised for freight containers are intended to simulate the dynamic loads occurring under normal conditions of transport. In undergoing the ISO tests, only elastic movement of the freight container structure is allowed. The ISO tests were considered to be more appropriate as representation of normal conditions of transport for freight containers, than the Type A tests. Also, freight containers designed in accordance with ISO 1496/1 have been proved, by the use of millions of units, to provide safe handling and transport under normal conditions of transport.

The 1990 Amendment to the IAEA Transport Regulations includes a containment criteria specific to para 523.

The following requirements for IP-2 freight containers, as defined in Paragraph 523 of the 1990 Amended issue of 1985 Edition of the IAEA Transport Regulations, are the basis for approval of IP-2 freight containers in the UK:

i) Packages must meet the general requirements for all packagings and packages.

ii) Packages must conform to the requirements prescribed in the International Standard ISO 1496/1-1978.

iii) When subject to the tests in ISO 1496/1-1978, packages must prevent:

   (a) The loss or dispersal of the radioactive contents; and

   (b) The loss of shielding integrity which would result in more than a 20% increase in the radiation level at any external surface of the package.

CONTAINMENT STANDARD

The containment standard specified for IP-2 packages (and for Type A packages) in the 1985 Edition of the IAEA Transport Regulations is that of 'no loss or dispersal' which has never been defined quantitatively in the regulations. The intent of the IAEA Transport Regulations in specifying a containment criteria for IP-2 packages, is to ensure that under normal conditions of transport the radioactive contents of the package cannot escape in sufficient quantities to create a radiological hazard.

In determining a practical and acceptable containment criteria for the IP-2 freight containers, due account was taken of the spirit of the IAEA Transport Regulations, guidance provided by the Advisory Material (Safety Series 37) [Ref 4], and the advice of the UK Competent Authority. The IAEA Transport Regulations provide for comparable levels of safety for radioactive materials of different radiotoxicities and different amounts, by relating the nature and amount of contents with graded packaging integrity requirements. The Explanatory Material (Safety Series 7 paragraph E-519.1) [Ref 5] states that "release of contents considerations for IP-2 packages impose a containment function on the packaging for normal conditions of transport". The Regulations also recognise that some simplification is possible with regard to containment standard for IP packagings, due to the nature of the LSA or SCO contents, as compared to the standard for Type A packagings that can contain an activity up to A_2 (not as special form radioactive material). The maximum activity of an IP package is limited by the package weight and allowable specific activity: for an ISO container carrying 20 tonne of LSA-III waste, the activity limit is 40 $\times 10^3$ A_2.

For small quantities of radioactive material (ie < A_2) within the Type A limits, that in slightly larger quantities would require a Type B package, the containment standard for the containment vessel is commonly taken to be that specified for normal conditions of transport of Type B packages. A Type B package can contain in excess of an activity of A_2 (not as special form radioactive material). This containment standard is usually demonstrated by a gas leaktightness test at all the verification stages: that is design, fabrication, assembly (pre-shipment) and periodic.

In the UK, it is accepted by the Competent Authority (Department of Transport) that solid particulate material would not be expected to leak from a seal having a gas leaktightness of 5 x 10^{-4} bar cm^3 s^{-1} SLR [Ref 6]. The acceptance, of this level of gas leaktightness does not apply to specially produced fine powders, but in practice no such radioactive powders are produced.

For Type A packages, leaktightness of the containment vessel is usually demonstrated at the design, fabrication and periodic stages only. Design verification is carried out by leakage tests before and after testing of prototypes. Assembly verification of the containment system after loading the radioactive content is assured by operational checks and controls, and normally no actual leakage tests are performed.

In determining a containment standard for IP-2 packagings the general approach described above for Type A packages was considered to be appropriate, but it was recognised that some simplification from that adopted for Type A packaging was possible within the spirit of the Regulations.

Various methods have been suggested of demonstrating containment following normal conditions of transport tests. These include visual inspection of the packaging, use of a suitable surrogate form of contents with visual inspection to assess contents retention (suitable surrogate materials being discrete objects or sand or flour as appropriate), and gas leakage testing.

QUALITY ASSURANCE

An important further consideration is that the 1985 Edition of the IAEA Transport Regulations emphasise the application of effective quality assurance and compliance assurance programmes to achieve safety of both the public and transport workers with respect to the transport of radioactive material.

In essence, the Regulatory requirements are directed at the shipper who is required to ensure that the package presented for transport will meet all the package design requirements, and specifically that:

a) the construction methods and materials used are in accordance with the design specification.

b) all packagings built to an approved design are periodically inspected, repaired and maintained in good condition so that they continue to comply with all relevant requirements and specifications, even after repeated use.

The 1985 Edition of the IAEA Transport Regulations advise that the Quality Assurance programmes should be commensurate with the complexity of the packaging and its components, and the degree of hazard associated with the contents that may be carried: to this effect a system of grading packages or components of packages is defined, where the grade relates to the safety significance of the package or component.

Detailed guidance on the graded approach to QA is given in the Advisory Material IAEA Safety Series 37 [Ref 4]. The following grades are defined:

GRADE 1 items — are those essential to safety and include items affecting package containment or shielding or criticality control.

GRADE 2 items — are structures or components or systems whose failure would indirectly affect safety.

GRADE 3 items — are those items whose failure would not affect safety.

The 1985 Edition of the IAEA Transport Regulations also advise as to the relationship between grading and package type. For IP-2 and IP-3 packagings, features affecting containment and shielding integrity are specified to be subject to GRADE 1. All other features should be subject to GRADE 2 except for those where there is a minimal effect on safety. It is noted therefore that the grading required for containment for an IP-2 package is the same as that required for Type A and Type B packages.

To meet the requirements of the 1985 Edition of the IAEA Transport Regulations, and in particular the Quality Assurance requirements, verification of the containment standard is required at the following stages:

- Design (Prototype Testing)
- Fabrication (Manufacture)
- Assembly (pre-shipment)
- Periodic (normally annually but may be longer when justified) for reusable containers

These stages of verification are being used for the IP-2 freight containers designed to meet the 1985 Edition of the IAEA Transport Regulations. These measures are seen as being commensurate with the appropriate grading (ie GRADE 1).

DEMONSTRATION OF CONTAINMENT PERFORMANCE

In determining an appropriate containment standard, and method of demonstrating this standard for IP-2 freight containers, a number of issues have been considered. The IAEA Transport Regulations advise that it is difficult to suggest a single test method due to the wide range of packagings and contents. However, it is suggested that a qualitative approach which involves testing, may be employed for IP and Type A packages. A method suggested for solid contents in particular, involves the measurement of pressure rise or drop under some type of vacuum or pressure test. A simple bubble test is suggested for gaseous contents.

In determining a design basis for IP-2 freight containers, and recognising that a qualitative approach must have a quantitative pass/fail criteria described, it was seen as appropriate that the containment standard should be expressed as a gas leakage rate and that this should be

verified at the appropriate stages. In determining an appropriate gas leakage rate the following factors were considered:

- what containment standard as appropriate for the form of the contents (ie LSA/SCO which affords a degree of inherent safety)

- what is practically achievable for large volume containment vessels

- what test and test sensitivity is achievable in relation to the proposed manufacturing methods

- some relaxation from the Type A (normal conditions for Type B) containment test criteria

In arriving at a practically achievable test method and containment standard expressed as a leakage rate, an extensive test programme was carried out involving gas leakage testing of prototype freight containers. A mass spectrometer was used to detect helium gas leaking from a pressurised freight container. Leak detection was carried out on the container body and seals by pressurising the container with helium gas and 'sniffing' the external surfaces of the container body to detect leaks. By using this technique it was found that a test sensitivity of around 10^{-1} bar cm^3 s^{-1} SLR could be reproducibly achieved for individual leaks for freight containers designed to be nominally leaktight. It should be noted that a mass spectrometer of intrinsic sensitivity of around 10^{-10} to 10^{-11} bar cm^3 s^{-1} SLR was employed, but the resultant sensitivity of about 10^{-1} bar cm^3 s^{-1} SLR was due to conductances in pipework, partial pressure of helium gas, background levels and the use of a 'sniffer' probe.

The resulting test sensitivity reflected the level which was achievable within the manufacturing environment and by constraints imposed by the nature of the freight container design (ie a thin walled flexible structure). Improvements in the test sensitivity could possibly have been achieved by removing the freight containers to a laboratory controlled environment. However, any technique developed under these conditions would not necessarily have been reproducible or practical in carrying out verification of containment standard during fabrication of large numbers of 'production' units.

This development work resulted in the conclusion that a containment standard of around 10^{-1} bar cm^3 s^{-1} SLR for individual leaks was achievable in the manufacturing environment. It was also concluded that this leaktightness could be reproduced, did not impose significant constraints on the design of IP-2 freight containers, and could, if required, be practically achieved and demonstrated at all stages of verification. Experience showed that all containment welds on the specially constructed freight containers could be readily leak tested to this standard. The development work provided useful information relating to the design of nominally leaktight freight containers, in particular, it was determined that it is necessary to ensure that all containment welds are accessible from outside the container and are visible.

As a further development, the sensitivity of a simple soap bubble test for measurement of leak rates on freight containers, was checked by a Helium leak detection technique. To achieve comparable results, both the Helium and Soap Bubble tests were carried out under the same test conditions. It was found that for a particular Helium sensitivity, as achieved in the tests described above, the soap bubble technique offered a comparable sensitivity: this was also demonstrated when pressurising the containers with air. The soap bubble technique using air as the test medium was thereafter adopted for the leak testing of all production units of IP-2 freight containers.

A gas leakage rate of 10^{-1} bar cm^3 s^{-1} SLR is equivalent to a single capillary of 40 micrometre diameter and of length of 2mm (equivalent to the freight container wall thickness). In comparison, it is noted that a capillary of 12 micrometer diameter and of 2mm length has a gas leakage rate of 5×10^{-4} bar cm^3 s^{-1} SLR: this leakage rate being that accepted in the UK as providing absolute containment for fine powders in relatively free form.

The smallness of the size of leakage path having a gas leakage rate of 10^{-1} bar cm^3 s^{-1} SLR was considered as unlikely to be the cause of powder leakage from freight containers because:

- there is virtually no pressure drive for leakage because the freight containers include filtered vents to ensure equalisation of internal pressure with ambient pressure. The filtered vents were included in the design because differential pressure changes were seen as undesirable, and difficult to design for, in a thin walled flexible structure.

- the radioactive material is not concentrated powders but powder mixed (diluted) with a spectrum of non-radioactive materials.

- the formation of aerosols containing radioactive material was considered to be unlikely due to the absence of significant mechanisms for making any radioactive powders airborne.

The closure systems of the IP-2 freight containers described above, are designed such that the containment standard of the closure seals can be verified independently of the container body. This is achieved by the use of a seal design that can be leak tested separately. The arrangement currently in use consists of two elastomeric seals separated by an interspace; a typical arrangement is shown in Figure 3. The resultant interspace volume can be pressurised and the pressure drop recorded over a period of time. This provides a measurement of the leakage rate of the seals. As this technique provides a measure of the gross leakage rate of the seals (ie not individual leaks), a relaxation of a factor of 10 was allowed resulting in an acceptance containment standard of 1 bar cm^3 s^{-1} SLR.

DESIGN FRAMEWORK

As a result of the above considerations, the design requirements listed below have been adopted in order to meet the performance requirements of the 1985 Edition of the IAEA Transport Regulations, for a freight container designated as an IP-2 packaging.

i) The freight container is to be designed to meet the design and test requirements of ISO 1496/1-1978.

ii) The containment standard is to be demonstrated at the Design, Fabrication and Periodic stages by simple gas leak tests.

iii) Resealable closures are to be designed with double elastomeric seals which allow gas leak testing of both the containment vessel and the closure seals at all appropriate stages.

i) For design verification, the containment standard for the containment vessel and seals is to be tested before and after the ISO tests (ISO 1496/1 testing). The containment standard of the closure seals is also to be verified during the ISO racking tests.

v) Assembly verification is to be provided by the combination of fabrication verification and written procedures. If repair of the closure is required after manufacture, the double seals enable leak tests to be carried out to ensure that the repair is satisfactory.

vi) For reusable containers, the containment standard of the seals is to be verified annually and a detailed inspection carried out on the serviceability of the container. The container is to be maintained every five years, including leak testing of the containment vessel and verification that the container meets the requirements of the ISO tests.

vii) The containment standard is to be demonstrated at the appropriate stages by a gas leak test. The leak tightness levels currently adopted are:

Containment Vessel: 10^{-1} bar cm^3 s^1 SLR (individual leaks detected)

Closure Seals: 1 bar cm^3 s^{-1} SLR (gross leaks detected)

DESIGN SPECIFICATION

For a freight container the above considerations lead to the following generic design specification to enable the appropriate IP-2 regulatory requirements to be met:

- containment vessel to be fabricated in steel and to be of a continuously seal welded construction with all seal welds accessible from the outside.

- closure (eg lid or door) to be fabricated from steel and be of a seal welded construction.

- elastomeric seals to be used between the containment vessel and closure.

- closure seal designed to be leak testable by the gas pressure drop method by use of double seals with an interspace.

- containment vessel (including lid or door) to be leak testable by soap bubble method.

- container to be designed to meet the design and test requirements of ISO 1496/1-1978.

- tiedown arrangements to be provided for contents.

- filtered vents to be provided to ensure equalisation of changes in pressure due to changes in ambient temperature and pressure.

CONCLUSION

Within the UK, freight containers are used in large numbers for the handling, transport, and disposal of LLW and to a lesser extent for other LSA or SCO materials. A design framework has been established for such freight containers to ensure that they meet the requirements of IP-2 packagings in accordance with the requirements of the 1985 Edition of the IAEA Transport Regulations.

The practicality of the interpretation of the Regulations embodied in this design framework has been proven by the design and manufacture of several different designs of IP-2 freight containers. The particular problem of providing a simple but adequate means of proving the adequacy of the containment of each freight container, at all the verification stages required by the Regulations, has been solved by establishing appropriate design criteria. The practicality of effecting this by testing has also been established on the large number of IP-2 freight containers built and tested.

The result of this work is that a cost effective engineering solution has been developed for the manufacture and operation of large freight containers for the transport of solid radioactive waste under the IP-2 packaging requirements of the 1985 IAEA Transport Regulations.

REFERENCES

Ref 1   IAEA Safety Series No 6, Regulations for the Safe Transport of Radioactive Material, 1985 Edition, and 1985 Edition (As Amended 1990).

Ref 2   IAEA Safety Series No 6, Regulations for the Safe Transport of Radioactive Materials, 1973 Revised Edition (As Amended).

Ref 3   International Standard ISO 1496/1, Series 1 Freight Containers - Specification and Testing - Part 1: General Cargo Containers, 1978.

Ref 4   IAEA Safety Series No 37, Advisory Material for the IAEA Regulations for the Safe Transport of Radioactive Material (1985 Edition), Third Edition (As Amended 1990).

Ref 5   IAEA Safety Series No 7, Explanatory Material for the IAEA Regulations for the Safe Transport of Radioactive Material (1985 Edition), Second Edition (As Amended 1990).

Ref 6   Private communication from Department of Transport to R A Vaughan, 21 December 1987.

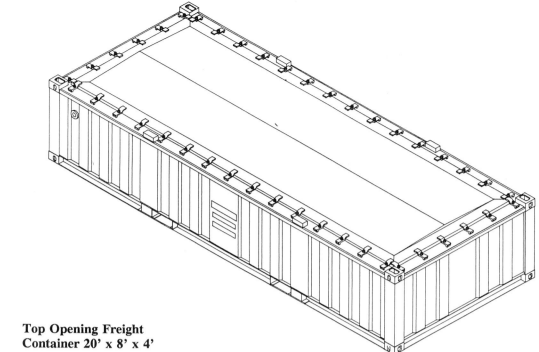

Fig 1   **Top Opening Freight**
            Container 20' x 8' x 4'

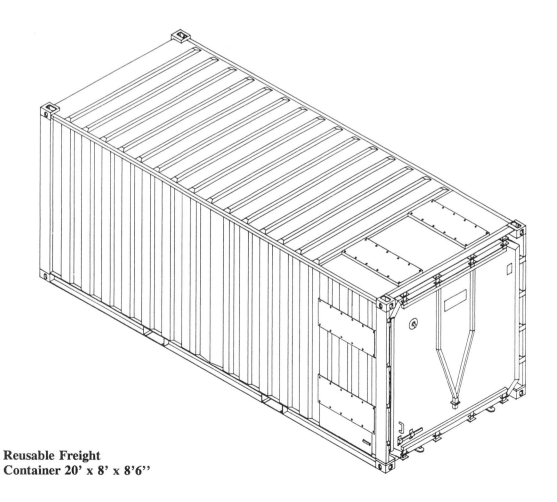

Fig 2   **Reusable Freight**
            Container 20' x 8' x 8'6''

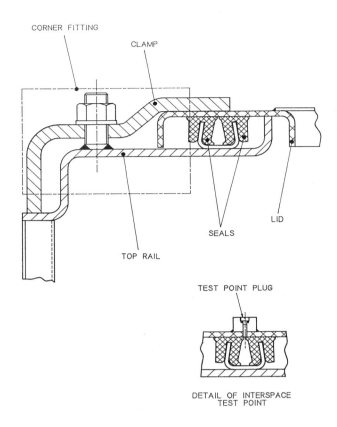

CORNER FITTING

CLAMP

LID

SEALS

TOP RAIL

TEST POINT PLUG

DETAIL OF INTERSPACE
TEST POINT

**Fig 3   Cross Section of a
Typical Seal Arrangement**

C431/017

Packaging for transport and disposal of low level waste at Drigg

S G HIGSON, BSc
British Nuclear Fuels plc, Risley, Warrington, Cheshire
R P HOWS,
Croft Associates Limited, Upton, Didcot, Oxfordshire

SYNOPSIS Solid low level waste (LLW) disposal operations at the British Nuclear Fuels plc (BNFL) Drigg site are currently being upgraded. A major feature of this upgrade is the introduction of waste compaction, containerisation and orderly emplacement of packages in concrete lined trenches (vaults). This paper summarises the current status of the upgrade with particular emphasis on progress towards specification of a product container design that is consistent with the overall aim of achieving long term post-closure site stability and will also meet the requirements for transport to Drigg through the public domain under the conditions of the 1985 IAEA Transport Regulations.

1   INTRODUCTION

British Nuclear Fuels plc (BNFL) owns and operates the principal UK solid low level waste (LLW) disposal site which is located at Drigg in West Cumbria. The site has a total area of 110 hectares of which 35 hectares are currently consented for disposal of LLW. Within this area there is an essentially continuous clay layer at about 5-8m depth. Past disposal operations involved the use of trenches cut so that the clay forms a graded, low permeability base. Infiltrating rainwater and groundwater is thereby directed to the southern end of the trenches where it is collected for discharge to the Irish Sea. Waste was tumble tipped into the trenches and covered with a layer of earth incorporating small boulders and geotextile sheet thus forming a flat stable surface from which further disposal operations could continue.

2   DRIGG UPGRADE

The above method of disposal was developed against a model of sanitary landfill practices and was efficient in terms of space utilisation, radiation dose uptake to workers and time taken for disposal operations. In recent years, however, it has been accepted that the appearance of the site and to some extent the management practices, particularly in terms of long term containment, should be upgraded. In 1987 therefore, BNFL announced a major programme of improvements to operations at Drigg including:

(a)   capping of completed trenches to limit rainwater infiltration;

(b)   refurbishment of the leachate drainage/discharge system;

(c)   containerisation of waste with prior compaction where appropriate;

(d)   orderly emplacement of containerised waste in concrete lined trenches (vaults).

A temporary cap has recently been installed over the completed trenches. This comprises a 1:25 graded earth mound incorporating a low density polyethylene membrane. A permanent cap incorporating a thick band of clay will be installed in the longer term once settlement is judged to be complete. Future vaults will be capped in a similar manner.

Refurbishment of the leachate drainage system was completed early in 1991 when new holding tanks and a marine pipeline were brought into operation for discharge of leachate direct to the Irish Sea rather than via local water courses.

Waste containerisation is being introduced on a phased basis. Waste from non-Sellafield consignors is now being routinely despatched to Drigg in purpose built ISO freight containers. Two basic types of containers are in use:

(a)   Full height end door containers (figure 1) with external dimensions 6.06m x 2.44m x 2.59m high. These containers are used for drummed waste, (each container holding about 90 nominal 200 1 drums) which is generally of relatively low density.

(b)   Half height, top lidded containers (figure 2) with external dimensions 6.06m x 2.43m x 1.22m high. These containers are used for waste which is too large (oversize) to load into drums and which is generally of relatively high density.

Both container designs incorporate four intermediate load transmitting legs (see figures 1 and 2) in addition to the conventional corner posts in order to reduce the magnitude of individual point loads at the base of each container stack.

Upon receipt at Drigg, the containers are stacked by fork lift truck in a new concrete vault (Vault 8). The vault is approximately 175m wide by 210-260m long and has an average depth of about 5.3m giving a total capacity of about 180 000m^3 of containerised waste.

Sellafield waste continues to be tumble tipped into the last open cut trench (Trench 7) on a temporary basis pending the provision of a purpose built compaction and containerisation plant at Sellafield.

Whilst waste compaction is already in use on a limited basis, plans for widespread application are still being developed. During 1989, a review of plans for compaction and containerisation of Sellafield waste (which then centred on low force in-container compaction into suitably strengthened full height ISO freight containers) indicated that residual voidage in ISO freight containers was likely to be significant even after the full introduction of compaction. This led to concern that extensive disruption of the water shedding vault cap could occur as the containers degraded and estimates indicated that the timescale for this was likely to exceed the credible period of site maintenance. It was therefore concluded that additional measures were needed in order to minimise residual voidage within the emplaced wasteform.

3 DRIGG WASTEFORM OPTIMISATION

Preliminary studies concluded that minimisation of residual voidage could best be achieved by the adoption of high force compaction in conjunction with in-fill grouting of readily accessible voidage once the resultant compacts (pucks) and any loose items that were not amenable to compaction had been loaded into the product container.

Conventional high force compaction requires prior packaging of waste generally in nominal 200 l drums. Whilst the infrastructure at non-Sellafield consignor sites is already compatible with this requirement, waste at Sellafield is currently collected loose in 5 and 10m³ skips for tumble tipping. Studies into the best means of pre-packaging Sellafield waste concluded that the optimum scheme involved provision of centralised facilities at the head end of the proposed compaction plant for:

(a) initial size reduction of oversize material in order to maximise the percentage of total arisings consigned to the compactor;

(b) loading waste into nominal 1m³ metal boxes rather than 200 l drums in order to minimise the extent to which initial size reduction would be necessary;

(c) in-box pre-compaction in order to minimise the requirement for primary containers.

The adoption of high force compaction and in-fill grouting will lead to increased product container masses and assessments indicated that the point loads resulting from continued use of the existing design of ISO freight container could compromise the integrity of the vault base slab. Further, the existing container design includes significant pockets of voidage in the base assembly and top (lid) area that would not be accessible to the in-fill grout. It was therefore concluded that the overall product container design philosophy should be reviewed.

Four basic container design schemes representing a range of extremes were considered:

(a) small concrete;
(b) large concrete;
(c) small metal;
(d) large metal.

Table 1 compares a number of key parameters for the four basic schemes:

Table 1   Comparison of Container Designs

	Relative Efficiency of Vault Utilisation	Relative % Residual Voidage	Relative No Handling Operations
Small Concrete	2.0	1.0	9.6
Large Concrete	1.8	1.0	1.4
Small Metal	1.0	1.9	10.8
Large Metal	1.1	1.6	1.0

From this table it can be seen that the efficiency of vault utilisation (ie the ratio of the external container volume to payload capacity) is better for the metal containers whereas the concrete containers result in lower percentage residual voidage. Both these effects are due to the relatively high bulk volume associated with the container structure (absolute voidage is in fact comparable for all four container schemes). Handling operations are obviously much reduced with the large overpacks which will have consequent benefits in terms of reduced operator dose uptakes.

In comparing the various schemes it was also noted that:

(a) metal containers would result in lower gross package masses prior to infill grouting;

(b) experience in the fabrication of concrete containers was limited;

(c) given the experience obtained with the existing range of ISO freight containers, there was greater confidence that metal overpacks could be shown to meet the requirements of the 1985 IAEA Transport Regulations.

It was therefore concluded that, on balance, the large metal overpack was the preferred scheme.

4   1m³ BOX DESIGN

As noted previously Sellafield waste is to be force compaction. The principal features of the conceptual box design are:

(a) simple construction from 16 SWG mild steel;

(b) biscuit tin type lid which is tack welded into place prior to compaction;

(c) overall dimensions (915mm x 1090mm x 1220mm high) optimised to suit product container dimensions;

(d) internal features to enable location of inner lids in order to prevent waste reassertion after in-box pre-compaction.

Figure 3 shows the prototype nominal $1m^3$ box.

5 DETAILED PRODUCT CONTAINER DESIGN

Three broad product container design objectives were specified:

(a) structural voidage that was not accessible to in-fill grout should be minimised;

(b) load distribution should be relatively uniform;

(c) the appropriate requirements of IAEA Safety Series 6 (Regulations for the Safe Transport of Radioactive Material, 1985 Edition) should be met.

In considering the design of the large metal product container, attention focused on the shortcomings of the existing designs of full and half height ISO freight containers. Handling and transport requirements indicated that the approximate dimensions of the existing design of half height ISO freight container were about optimum, resulting in an average gross grouted product container mass of about 30t. The height of the container was, however, increased slightly (from 1.22m to 1.3m) in order to maximise the efficiency of vault utilisation.

The main problems relating to the existing design of half height ISO freight container were:

(a) The overall voidage associated with the container structure that cannot be readily in-fill grouted is about $3.5m^3$ (about 20% of the external envelope of the container). The majority of this voidage is in the base assembly which comprises a flat plate floor supported on cross beams (the floor prevents grout access to the void spaces between the cross members and, in any case, the bottom of the base assembly is unenclosed). The remainder of the voidage arises due to the corner posts and intermediate legs which protrude above the main container body to enable stacking without prejudice to the container lid (see figure 2).

(b) The use of legs and corner posts for load transmission is not compatible with achieving relatively uniform load distribution.

In order to achieve relatively uniform load distribution it was concluded that container stack loads would have to be transmitted through the wasteform rather than the container structure. This was judged to be readily achievable since the compressive strengths of the waste after high force compaction and the in-fill grout would both be about 10^4 kNm^{-2} compared to a maximum container plus cap load of about 10^2 kNm^{-2}. Transmission of load via the wasteform required good contact between stacked containers across the majority of the plan area implying a need for flush, flat tops and bases to the containers with no protrusions and good contact between these faces and the in-fill grout.

The original intention had been to incorporate an in situ cast grout lid in order to form the flat top to the completed package. It was noted, however, that:

(a) puck density was likely to be less than the in-fill grout density giving rise to puck flotation problems during in-fill grouting;

(b) the permeability of the grout may lead to difficulties in demonstrating compliance with the 1985 IAEA Transport Regulations requirement for 'no loss or dispersal'.

It was therefore concluded that a sealed metal inner lid should be adopted for containment purposes. Whilst such a lid would also prevent problems associated with puck flotation, it was noted that the large panel area of the lid may be susceptible to distortion under the pressures of grouting and thus an outer in situ cast grout lid was also proposed to ensure a stable flat surface to the completed package.

A prototype container designed to meet the requirements of relatively uniform load distribution, minimum structural voidage and compliance with the 1985 IAEA Transport Regulations has been manufactured. Figures 4-6 illustrate the principal features.

Figure 4 shows that a solid base has been provided across the bottom face of the base cross members (for manufacturing simplicity, the bases of the fork pockets have not been enclosed).

Figure 5 shows the inside of the container and in particular the provision of slats above the base cross members thus permitting grout access whilst maintaining the structural strength of the base assembly.

Figure 6 shows the container with the sealed metal inner lid in place. The principal features of the lid are:

(a) 'Well type' construction to form a mould for the in situ cast grout lid.

(b) Extension of the side walls of the 'well' above the level of the container body side walls which, in conjunction with the container body side walls and top rail forms a gutter around the top perimeter of the container. This arrangement allows grouting to proceed to the point of overflow with any surplus grout collecting in the gutter. It also guarantees that the level of the grout surface is above the level of the corner posts thus ensuring that the load is transmitted through the wasteform rather than the container structure.

(c) A 'weld mesh' grid to key the capping grout to the lid.

(d) Ports with removable sealed cover plates for grout access and venting of displaced air (Note: the precise number and location of ports is still being optimised in conjunction with grouting plant design).

All aspects of the container design have been assessed against the requirements of the 1985 IAEA Transport Regulations. In order to meet the regulations the container is designed and tested to meet the requirements of the ISO standard (ISO 1496-1978, Series 1 Freight Containers Specification and Testing-Part 1 : General Cargo Containers) with the additional requirement of 'no loss or dispersal' of the contents under normal conditions of transport as represented by the ISO tests. In order to demonstrate this the container has been tested against the requirements of the "Convention for Safe Containers" (CSC) and in addition leak tested to 10^{-1} bar cm^3 s^{-1} SLR (AECP 1068, Leakage Tests on Packages for Transport of Radioactive Materials (Provisional), April 1988).

6   COMPACTION PLANT DESIGN

Detailed design of a Waste Monitoring and Compaction (WAMAC) facility and associated grouting facility for processing Sellafield waste is now well underway with an estimated date for start of operations of late 1993.

The main feed to WAMAC will be loose Sellafield LLW in 5 and $10m^3$ skips. Upon receipt, these will be emptied (tipped) and the contents loaded remotely into nominal $1m^3$ metal boxes. Oversize waste will be size reduced by either shredding or using a remote control vehicle with interchangeable cutting tools. The waste will be in-box pre-compacted and, when the boxes are full, lids will be tack welded into place. The boxes will then be high force compacted and the resultant pucks transferred to the product container by suction lift. Finally, the metal inner lid will be fitted prior to shipment to the grouting facility for in-fill grouting and application of the in situ cast grout outer lid.

Similar measures are proposed for non-Sellafield consignors who will either make use of the facilities at Sellafield or make alternative arrangements for high force compaction and packaging at source. Whilst there is the option for non-Sellafield consignors to in-fill grout at source, transport considerations suggest that grouting will generally take place in the BNFL grouting facility which it is envisaged will be sited at Drigg.

7   CONCLUSION

Concerns regarding long term Drigg site stability have resulted in the specification of a revised wasteform. This involves the introduction of high force compaction of all suitable waste and in-fill grouting into a top lidded half height ISO freight container specially designed to ensure relatively uniform load distribution when stacked in the vault at Drigg. Detailed container design has been completed and a prototype has been manufactured and tested to demonstrate compliance with the 1985 IAEA Transport Regulations.

Fig 1 Full Height ISO Freight Container

Fig 2 Half Height ISO Freight Container

Fig 3 Prototype 1m^3 Box

Fig 4 Modified Half Height ISO Freight Container:
Base Detail

Fig 5 Modified Half Height ISO Freight Container:
Inner Detail

Fig 6 Modified Half Height ISO Freight Container:
General Arrangement

C431/044

The Windscale AGR intermediate level decommissioning waste management system

J H TRATT, BSc, CEng, MIEE
AEA Technology, Windscale, Seascale, Cumbria

SYNOPSIS

The Windscale Advanced Gas-Cooled Reactor (WAGR) was shut down in 1981 after 18 years operation at high load factor. It is currently being dismantled to achieve Stage 3 decommissioning, ie complete clearance of the site. These operations will give rise to several hundred tonnes of Intermediate Level Waste over the next five years. This material requires to be handled, shielded, encapsulated and stored until the UK NIREX Ltd (Nirex) deep repository is available. The engineering solutions are described, together with the requirements of the statutory organisations.

1 INTRODUCTION

The Windscale AGR (WAGR) was shut down in 1981 after 18 years of operation. The plant is now being used to develop methods for decommissioning and to demonstrate the feasibility of ultimately decommissioning to a green field site.

The reactor has been completely defuelled and the remaining radioactive material consists mainly of structural materials such as steel, graphite and concrete, which have become activated by neutron irradiation. (Figure 1).

The reactor standpipes and pressure vessel top dome have been dismantled down to the level of the hot box using 'hands-on' techniques and the waste generated, being low level, has been disposed of at BNFL's Drigg surface repository.

Further dismantling of the reactor will be carried out using remote handling techniques and will generate approximately 900m³ of Intermediate Level Waste (ILW). The handling, packaging and storage of the ILW is the subject of this paper.

2 NATURE OF THE WASTE

When the reactor was shut down in 1981 a complete radioactive inventory was prepared using calculations based on records of neutron flux throughout the operating period and of the materials of construction. It was estimated that the total activity of the plant was 7.4×10^{15}Bq. Since that time, samples have been taken from the more accessible components of the reactor, enabling the accuracy of the calculational methods to be checked and improved. The current estimate of activity of ILW in 1993 is 2.90×10^{15}Bq. This is the year that remote dismantling is planned to commence. Natural decay will reduce this to 5.14×10^{14}Bq by 2005 and to 1.66×10^{14}Bq by 2055. The marked reduction of activity over the next 50 years is largely due to the decay of Co60 and Fe55. ILW activity is shown by material in Table 1 and by radionuclide in Table 2. Operational experience indicates that, with the exception of the four heat exchangers, contamination levels will be insignificant compared with activation levels.

In terms of weight, the most significant components of the

reactor are the graphite core (210 tonnes) and the thermal shield plates (186 tonnes). The thermal shield is also the most highly activated mild steel in the reactor, contributing 1.06×10^{15}Bq (50.7%) of the 1993 total activity. These, and most other components, are of a rectangular form, or can be collected in rectangular baskets. This influences the choice of the shape chosen for the waste package.

3 THE WASTE ROUTE

Calculations of activity levels led to a decision that all dismantling work following removal of the reactor pressure vessel top dome should be done by remote operation. Methods of remote cutting and segregation of the various reactor components have been developed, together with a system for transporting the pieces out of the reactor vault into a purpose built Packaging Building. (Figure 2).

A shielded route has been provided between the vault and the Packaging Building by jacking up two of the heat exchangers and making use of their concrete bioshields. One of the spaces thus provided has been developed into a connecting passage into the packaging building and the other is used as a maintenance cell for contaminated dismantling equipment. (Figure 3).

Reactor components are moved from the vault into the Sentencing Cell of the Packaging Building by means of a remotely controlled hoist. A material sampling station is built into the Sentencing Cell to allow samples to be taken from selected components and posted out for off-line analysis.

In the Sentencing Cell, components are put into the required configuration for placing into the waste container. This operation may involve rotation of cut plates from the vertical to the horizontal plane, or placing of components in specially designed baskets, which are then placed in the waste container.

Immediately below the Sentencing Cell is the Upper Loading Cell. This cell accommodates the equipment used for on-line assay of steel components,

which is dealt with in the next section of this paper.

Below the Upper Loading Cell lies the Lower Loading Cell and it is here that the waste is placed in the container and the full container is monitored to ensure that it complies with the LSA III regulatory limits on radiation. It is possible to remove waste from the container, if necessary, to obtain such compliance.

The loaded container is then moved to the Concreting Cell, where the contents are immobilised by grouting, and, after curing of the grout, a reinforced concrete closure is cast.

The container may then be removed from the Concreting Cell, washed down and monitored for external contamination and placed in the weighing pit. It remains here while the closure is cured and is then prepared for transportation to the Waste Container Store.

4 ASSAY AND DATA RECORDING

A major Quality Assurance requirement of the waste management system, is the knowledge of the contents of each waste container. Those radionuclides which are considered to be important to handling, transport and disposal in a deep repository have been identified and systems have been developed to record the quantities of these nuclides deposited in each waste container.

4.1 Assay

Provision for the assay of wastes from WAGR has been engineered into the Waste Packaging Building with the incorporation of ion chambers and the provision of a specially built handling cell, the Materials Sampling Station.

Due to practical difficulties of measuring beta emitters, especially on-line, gamma emitters alone will be monitored directly. A series of computer codes, is being developed to convert ion chamber readings into bulk Co60 activity levels. Off-line analyses of samples will be made for the elemental cobalt content of wastes to determine the neutron irradiation (flux and time) the item experienced. With this data the activity of other radionuclides will be estimated by computer codes.

Computer codes for the calculation of bulk gamma activities will be checked for accuracy against hand calculations and other codes with a verifiable QA route. Computer codes for the calculation of bulk beta activities will be checked for accuracy against hand calculations. The calculational route will also be validated by off-line analysis of material samples.

The proposed operational methodology is that:

a. Items of steel waste removed from the reactor structure will be monitored using ion chambers in the Upper Loading Cell of the Waste Packaging Building. These measurements will be used to calculate a total Co60 activity for the waste using a specially developed computer code. The waste item will then be transferred to the Materials Sampling Station, for spectrometry measurement of Cs137 (and possibly Nb94 for stainless steel). Samples will then be removed for off-line analysis of elemental cobalt. The neutron flux experienced will be calculated from the Co59/Co60 ratio, allowing the activities of Fe55, Ni59, Ni63 and Nb94 and hence total activity to be calculated.

b. Items of graphite, concrete and insulation will be monitored by the removal of samples for off-line analysis in the Materials Sampling Station of elemental cobalt levels. Co60 levels will be determined on-line in the Materials Sampling Station, but levels of the beta emitters will be estimated by calculation as described above.

Table 3 summarises the analytical methods proposed to determine the major isotopes.

α-emitters will not be assayed. They are currently estimated at 10^{-4} of βγ. This ratio has been determined by direct measurement of contamination in the heat exchangers and is considered pessimistic in terms of WAGR waste in general.

4.2 Data Recording

The principal element of the data recording process is a computerised data logging system with appropriate applications software. Each item of waste will be assigned a unique alpha numeric identifier and the following information will be recorded:

* Item description.

* Position within the reactor structure.

* Weight.

* Size.

* Source/detector distances (in Upper Loading Cell).

* Ion chamber dose rates.

* Isotopes/specific activities.

* Total activity.

* Time and date.

* Material composition.

Records relating to waste composition and activity will be transferred to the Nirex waste package database, before the waste boxes are despatched to the Nirex deep repository.

4.3 Sample Archiving

During the course of the assay process, samples will be removed from waste items for long-term storage in an archive. The objectives are:

* To provide a referable record of materials which have passed through the waste route.

* To provide a database of materials typically used in the construction of a gas cooled reactor.

* To provide reference data via measurement of samples to allow on-line assay equipment to be calibrated.

The frequency of sampling will be dependent on the nature of the waste items, material type and irradiation history and the data will be recorded in conjunction with the unique waste identifier.

5 WASTE CONTAINER DESIGN

5.1 Shape and Size

As previously stated, the majority of

the waste to be packaged will be of rectangular form. For efficient packing, the internal shape of the container should conform to this pattern, and, if a regular shielding thickness is to be provided, so should the external profile.

At the time that development of a suitable container was initiated, no size standards had been set by Nirex. Indeed, at that time Nirex had not been established and dumping at sea was the only available disposal option, and design was governed by the London Convention on Sea Dumping, (Reference 1), and the IAEA Regulations for the Transport of Radioactive Material (Safety Series 6, 1973). On the assumption that the containers would at some time travel by rail, external dimensions were determined by British Rail loading gauge W5 in conjunction with a currently available eight-axle truck. The overall weight of a loaded container was set at 50 te maximum. Subsequently, sea dumping was ruled out in favour of burial in a deep repository, and, following discussions with Nirex, the container size was set at 2438 x 2210 x 2200mm high. The current container design is illustrated in Figure 4.

5.2 Material of Construction

The container is required to provide strength, containment and shielding, in accordance with the relevant transport regulations. In the case of sea dumping, it must retain these features during the descent to the seabed. For land burial, the container forms one of a series of barriers providing containment. Taking these factors into account, along with economic considerations, reinforced concrete was considered the optimum choice of material. It is cheap, plentiful and strong, and requires no new technology. It is also available in a range of densities, which is valuable when considering the shielding requirements for some of the more active reactor components, as it allows the container dimensions to be kept constant without a significant decrease in payload.

5.3 Functional Requirements

The Functional Specification for

the container, addresses the requirements imposed by regulations covering its transportation and eventual disposal, as follows:

5.3.1 Functional Requirements Imposed by the Transport Regulations

5.3.1.1 The container design is governed by the following documents:

* IAEA Safety Series No. 6: Regulations for the Safe Transport of Radioactive Material, 1985 Edition. (As amended 1990).

* IAEA Safety Series No. 7: Explanatory Material for the IAEA Regulations for the Safe Transport of Radioactive Material (2nd Edition). (As amended 1990).

* IAEA Safety Series No. 37: Advisory Material for the IAEA Regulations for the Safe Transport of Radioactive Material, (1985 Edition), Third Edition.

* IAEA Safety Series No. 80: Schedule of Requirements for the Transport of Specified Types of Radioactive Material Consignments. (As amended 1990).

* AECP 1076: A Guide to Radioactive Material Package Design and Approval Requirements.

* Ionising Radiation Regulations, SI 1985 No. 1333.

* BR 22426: List of Dangerous Goods (L D G) and Conditions of Acceptance by Freight Train and by Passenger Train.

Whilst designing to the current Regulations it is appreciated that the regulations may change by the time that transport is necessary. It may be assumed that radioactive decay will mean that lower dose rate limits will be achievable in the future.

5.3.1.2 Waste classification

At the time of reactor dismantling, the majority of the WAGR radioactive decommissioning intermediate level waste is expected to fall within the

Low Specific Activity Material Category (LSA II or LSA III) as defined in paragraph [131] of the 1985 Edition of the Transport Regulations.

Approval will be sought from the Department of Transport for categorisation of the WAGR waste as either LSA-II or LSA-III as most appropriate ([A-131], Safety Series No. 37). This will be demonstrated by calculation of the specific activity of items to be packaged based on the most up-to-date activation calculations.

For the LSA-III category, the radioactive material should be relatively insoluble, or intrinsically contained in a relatively insoluble matrix. It will be necessary to demonstrate that if the entire contents of a package were subjected to the test specified in paragraph [603], the activity in the water would not exceed $0.1\ A_2$. This will be assessed by calculation, as the bulk of the material will be encased in cementitious grout and will be only lightly contaminated.

Packages will be made up such that the external radiation level at 3 metres from the unshielded material does not exceed 10 mSv/h [422]. This will be demonstrated initially by calculation and assessed by direct measurement before consignment to the Waste Container Store ([A-422], Safety Series No. 37).

Since the filled containers will be transported under exclusive use conditions, low specific activity material in both LSA-II and LSA-III categories may be packaged in containers which meet the design requirements of industrial packages Type 2 (IP-2). These requirements are set out below for convenience but reference should also be made to the 1985 Transport Regulations (relevant paragraphs shown in square brackets).

5.3.1.3 Design requirements for Type 2 industrial packages (IP-2)

The container will be designed to meet the general requirements for all packagings and packages [426, 505-514].

Air transportation [515-517]

will not be used.

The container, when subjected to the test specified in paragraph [622], 0.3m drop test, will prevent:-

a.   The loss or dispersal of the radioactive contents; and

b.   The loss of shielding integrity which would result in more than a 20% increase in the radiation level at any external surface of the package.

This may be demonstrated by reviewing the drop tests already carried out but the possibility of further tests is not excluded.

The designer will show that the requirements for the stacking test (5 times the package mass) given in paragraph [623] can be met.

5.3.1.4 Shielding/radiation levels

Container wall thicknesses will be such that the radiation levels do not exceed 2 mSv/h at any point on the external surface and 0.1 mSv/h at any point 2 metres from the outer surface [469]. However, the completed packages must also meet the NIREX requirements for dose rates when taken to the Repository (see Section 5.3.2.3).

The construction of the walls, base and lid will be such that the shielding properties and integrity are maintained as required by these regulations.

5.3.1.5 Contamination

The container outer surface must have the potential to be decontaminable so that the limits given in the Transport Regulations can be met, ie the non-fixed contamination on the external surfaces of the container will be kept as low as practicable and, under conditions likely to be encountered in routine transport, will not exceed 4 Bq/cm^2 $\beta\gamma$ activity when averaged over any area of 300 cm^2 [408].

5.3.1.6 Tie-down system

A satisfactory tie-down system or means of restraint will be required for all modes of transport. There is a possibility of using twistlock pockets in the base.

5.3.1.7 Marking and labelling

Provision should be made for labelling of packages prior to transport as required in paragraphs [436, 440, 441, 442] of the 1985 Transport Regulations.

5.3.2 Functional Requirements Imposed by UK NIREX

The WAGR container is not a standard Nirex package, and therefore Nirex requirements for the container could not be specified exactly: Specifications for standard packages, such as 500 l drums and the $3m^3$ box, were therefore used as the basis for specifying functional requirements.

5.3.2.1 Design life

The design life of the container will be such that the package will be suitable for safe transport and handling after an on-site storage period of up to 50 years. The design of the container shall also provide containment for a minimum period of 50 years post emplacement in the repository.

5.3.2.2 Activity content

The activity content will be that of ILW, ie no wastes producing significant heat.

5.3.2.3 Dose rate

The dose rate will not exceed 0.1 mSv/h at one metre from the package surface, when taken to the Repository.

5.3.2.4 Heat output

The average heat output from all waste will not exceed 4 watts/m^3. The total heat output from a single container will not exceed 500 watts.

5.3.2.5 Surface contamination

The limits on surface contamination are as defined for the Transport Regulations (Section 5.3.1.5).

5.3.2.6 Lifting features

The container will be provided with twistock fittings for lifting and handling. The lifting features will have adequate factors of safety to withstand snatching forces. The container will be

sufficiently robust and durable to be bottom lifted for the first 50 years post emplacement in the repository.

5.3.2.7 Weight

The weight of the container will not exceed 50 tonnes.

5.3.2.8 Organics

The level of organics of the total container and its contents will be as low as reasonably practicable.

5.3.2.9 Venting

There will be only minimal gas generation within the packages and any build-up of pressure will be prevented by diffusion through the concrete walls of the container.

5.3.2.10 Voids

All reasonable measures will be taken to minimise the presence of voidage within the package.

5.3.2.11 Impact strength

Specific design measures will be taken to minimise localised damage such as spalling of concrete from the container surface which would otherwise be expected during normal transport and handling operations.

The waste package will be capable of being dropped in any attitude from a height of 0.3m onto an unyielding flat surface whilst retaining its radioactive contents and still remain suitable for safe handling, transport and disposal.

The release of activity in the form of particulate material will be estimated for impacts from up to 25m.

5.3.2.12 Stackability

The waste container will be capable of withstanding stacking loads resulting from stacking to a minimum of six high. Consideration will also be given of the ability of the container to stack up to a maximum height of 35m.

5.3.2.13 Identification

Each container will be marked with a unique ten character identifier as specified by the Nirex waste package identifier specification. The identifier will be marked on each of the four vertical faces. The characters will be typically 6-10 mm high and will be capable of being machine and man read after the fifty year atmospheric storage period and transport to the Repository. Also, they will be capable after being read by some means after a minimum of 50 years post disposal.

5.3.2.14 Permeability and leach rates

Concrete used in the containers will have as low a permeability as is reasonably practicable. Due consideration will be given by the designer to the control of crack widths which will be as small as is reasonably practicable and be within the range which can be considered to be self-healing.

The integrity of the container will be such that it will provide containment post emplacement at the waste Repository for a period of at least 50 years.

The container will, after placing and grouting into final position in the Repository, withstand hydrostatic pressures associated with 1000m head of water.

5.3.2.15 Fire resistance

When choosing the types of concrete and aggregate for the container, the vulnerability of the material to fire within the Repository may be, depending on the design of the repository, a significant factor. The design of the repository is still evolving, and therefore on a design basis the box designers will consider its behaviour in a 800°C fire lasting for 30 minutes.

5.3.3 Functional Requirements Imposed by British Rail

The following British Rail documents are relevant:

British Rail Loading Gauge W6.

BR 22426: List of Dangerous Goods (L D G) and Conditions of Acceptance by Freight Train and by Passenger Train.

5.3.3.1 The designer will demonstrate that adequate means of restraining the packages to suit BR requirements are possible within the W6 gauge.

5.3.3.2 The Authority Transport Liaison Officer will seek a letter from BR confirming BR have no objections to the handling and transportation of the packages proposed.

5.3.4 Functional Requirements Imposed by Sea Dumping

This section of the specification is to be read in conjunction with the following documents:-

IAEA Definitions associated with the London Convention.

MAFF Code of Practice DSCP1.

5.3.4.1 The total activity averaged over 1000 tonnes of waste plus container will not exceed 3.7 TBq/te (100 Ci/te).

5.3.4.2 The strength of the loaded container will be capable of withstanding an external pressure equivalent to 6000 metres head of water to maintain the sea dumping option. The container will also be sufficiently robust to arrive intact at the sea bed at a depth of 4000 metres at the disposal site. The specific gravity will be not less than 1.2.

5.3.4.3 Final approval of the authorising department, MAFF, would have to be sought whenever sea dumping restrictions were lifted.

5.4 Infill Material

The IAEA Transport Regulations require that the radioactive material be encapsulated within solid matrix. This has several functions. It fixes any surface contamination that may be present; it fills voids and spaces within the container and thus resists implosion; it secures the loose components and reduces internal damage and thus adds considerably to the crushing and impact strength; it provides and reduces the movement of water towards and away from the radioactive contents. Some of these features are present by design but some may be regarded as enhancements to the concept. A further example of the latter category is the beneficial influence of the infill material on the chemistry of the steel waste under the repository conditions.

The infill material selected

is a self-levelling grout prepared from Pulverised Fuel Ash (PFA) and Ordinary Portland Cement (OPC), with a water/solids ratio of 0.42. The grout is batch mixed using a high shear mixer and pumped into the container.

Tests using simulated waste in half scale containers, have demonstrated that the grout penetrates and fills gaps <1mm between steel plates. The measured voidage was 0.03% of the box volume.

After a drop test from a height of 5m, no major faults or cracks were visible in the steel waste/grout matrix. Three hair-line cracks, approximately 60mm in length, were seen in the region between the waste and the container wall.

5.5 Development Programme

The development programme commenced in 1983 and the work carried out in the period 1983-1988 has been reported in detail by Wakefield. (Reference 2).

More recently, a further series of drop tests from a height of 5m has been carried out on half-scale containers. The results of this series of tests have validated a computer code which has been developed to predict the impact response of the container. This will enable the effect of design changes, for example, the inclusion of Twistlock fittings, to be assessed without the need for further expensive and time consuming test programmes, unless the analysis indicates a potential major design fault. In this case it may be desirable to confirm the results of the analysis by testing.

The inclusion of Twistlocks, both for lifting and for hold down during transportation, has required development of suitably rated fittings and additional reinforcement of the box to sustain the lateral load tests required by BS 3951: Part 2.

The container design is now in the process of approval by the relevant regulatory bodies and it is possible that further development work will be required to allow any comments to be addressed.

6 WASTE CONTAINER STORE

The WAGR decommissioning project will commence generating Intermediate Level Waste early in 1993. Since this is many years ahead of the date when a deep repository will be available, arrangements have to be made for safe interim storage on site.

Several studies were carried out into methods of storing the containers over a prolonged period. Options examined ranged from simple hardstanding with a distance barrier, to a fully shielded structure with an internal travelling crane. A review of these options recommended that a shield wall store and an independent mobile transporter be adopted. The store will accommodate 71 containers at floor level, with a clear height to allow stacking 3-high. This allows a 17.0m wide manoeuvring aisle for the transporter and spacing of 200mm and 1000mm between ranks and files respectively. It also allows space for temporary moving and re-stacking of boxes, in order to gain access to any one box for inspection, should this ever prove necessary.

The building is designed for a 50 year life and construction materials have been chosen to minimise maintenance. An extract ventilation system will be provided, controlled so that it is only operational when the access doors are open, to disperse exhaust fumes from the diesel powered transporter. The building will be unheated.

The proposed transporter is a diesel powered fork lift vehicle of 50 te capacity, which will lift and transport containers by means of a purpose built lifting frame engaging the top twistlocks.

The choice of self-shielded containers permits the use of a lightly shielded enclosure. The 9m high shield wall design will limit the external dose rate to 2.5µSv/hr. Entrance to the store will be via a shielded access corridor to minimise radiation shine outside the building. Lead glass shielding around the driver's cab of the transporter will reduce the dose to the driver, from a waste container on the transporter, to 55 µSv/hr.

7 ACKNOWLEDGEMENT

The work reported was undertaken as part of the WAGR Decommissioning Project, funded by the UK Department of Energy.

8 REFERENCES

1. Reports of the London Convention on Sea Dumping of Radioactive Wastes. 1984-85.

2. WAKEFIELD J R. The Development and Testing of a Container for the Transport of Decommissioning Wastes. International Conference on Transportation for the Nuclear Industry - Stratford, May 1988.

TABLE 1

WAGR TOTAL ILW ACTIVITY BY MATERIAL AT 1993

	Mass tonnes	Activity Bq 1993	% Total Activity 1993	% Total Weight
Mild Steel	535	2.14E+15	73.7	45.4
Stainless Steel	28	7.06E+14	24.3	2.4
Graphite	210	3.70E+13	1.3	17.8
Concrete	338	1.13E+13	0.4	28.7
Insulation	67	7.93E+12	0.3	5.7
Total	1178	2.90E+15		

TABLE 2

WAGR TOTAL ILW RADIONUCLIDE ACTIVITY AT 1993

Nuclide	Bq 1993
H3	4.66E+13
C14	3.30E+12
C136	1.41E+11
K40	1.21E+09
Ca41	5.35E+10
Mn54	9.10E+08
Fe55	2.03E+15
Co60	5.68E+14
Ni59	2.34E+12
Ni63	2.43E+14
Nb93m	8.90E+10
Nb94	7.55E+10
Cs137	1.20E+11
Eu152	1.25E+12
Eu154	1.09E+12
Eu155	2.96E+11
Alpha	1.40E+07
Total	2.90E+15

TABLE 3

DETERMINATION OF MAJOR ISOTOPES

Material	Isotope	Method of Determination
Steel	Ni-59	Chemical detn. of total Ni followed by calc. of Ni-59
	Co-60	Gamma spectrometry for Co-60 Chemical detn. of total Co to give Co-59/Co-60 ratio
	Nb-94	Chemical detn. of total Nb followed by calc. of Nb-94
	Cs-137	Gamma spectrometry for Cs-137
Concrete/Graphite		
	H-3	Extraction/Separation followed by liquid scintillation
	C-14	Extraction/Separation followed by liquid scintillation
	Cl-36	Separation by ion chromatography followed by liquid scintillation
	Ca-41	Chemical detn. of total Ca followed by calc. of Ca-41 or Separation by ion chromatography followed by liquid scintillation
	Co-60	Gamma spectrometry for CO-60 Chemical detn. of total Co to give Co-59/Co-60 ratio

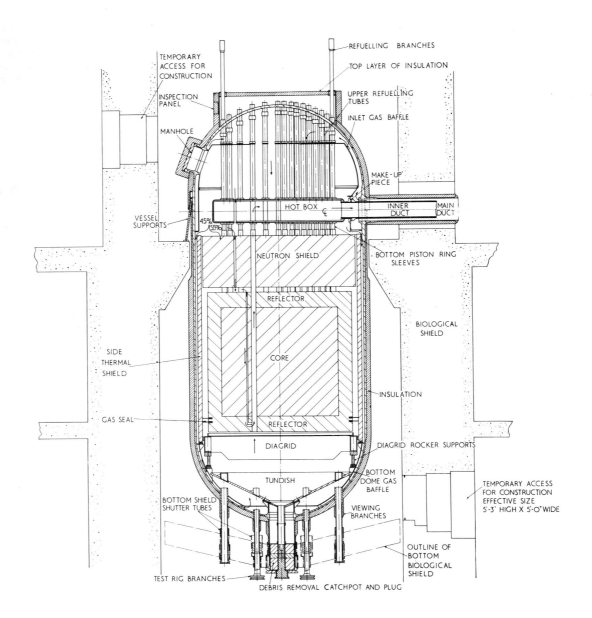

TEMPORARY
ACCESS FOR
CONSTRUCTION

INSPECTION
PANEL

MANHOLE

VESSEL
SUPPORTS

SIDE
THERMAL
SHIELD

GAS SEAL

BOTTOM SHIELD
SHUTTER TUBES

TEST RIG BRANCHES

REFUELLING BRANCHES

TOP LAYER OF INSULATION

UPPER REFUELLING
TUBES

INLET GAS BAFFLE

MAKE-UP
PIECE

HOT BOX

45%
55%

INNER
DUCT

MAIN
DUCT

NEUTRON SHIELD

BOTTOM PISTON RING
SLEEVES

REFLECTOR

BIOLOGICAL
SHIELD

CORE

INSULATION

REFLECTOR

DIAGRID

DIAGRID ROCKER SUPPORTS

TUNDISH

BOTTOM
DOME GAS
BAFFLE

VIEWING
BRANCHES

TEMPORARY ACCESS
FOR CONSTRUCTION
EFFECTIVE SIZE
5'-3" HIGH X 5'-0" WIDE

OUTLINE OF
BOTTOM
BIOLOGICAL
SHIELD

DEBRIS REMOVAL CATCHPOT AND PLUG

FIGURE 1   REACTOR VESSEL INTERNALS

C431/044

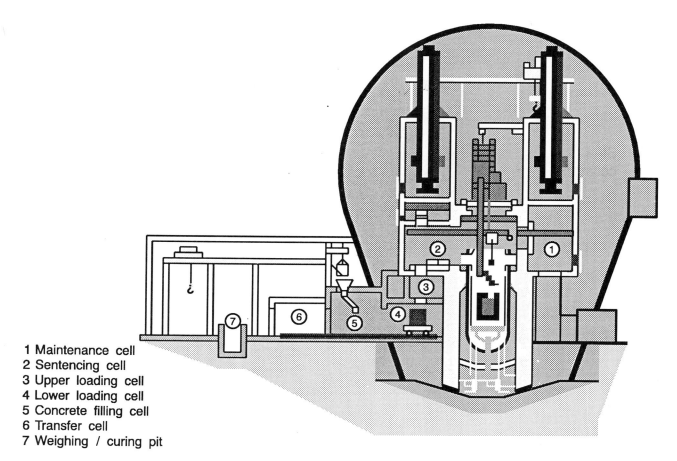

1 Maintenance cell
2 Sentencing cell
3 Upper loading cell
4 Lower loading cell
5 Concrete filling cell
6 Transfer cell
7 Weighing / curing pit

FIGURE 2   THE WASTE ROUTE

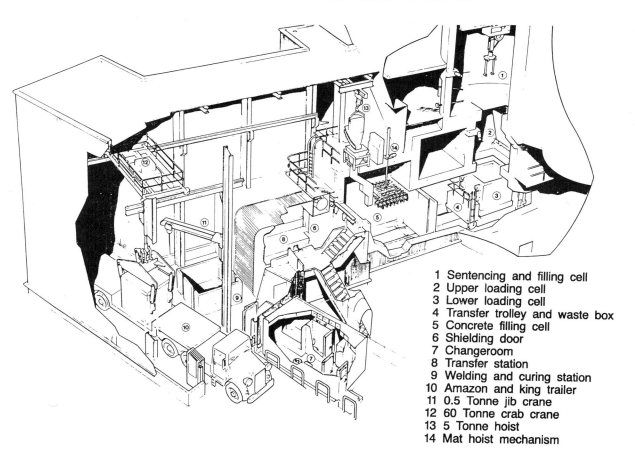

1 Sentencing and filling cell
2 Upper loading cell
3 Lower loading cell
4 Transfer trolley and waste box
5 Concrete filling cell
6 Shielding door
7 Changeroom
8 Transfer station
9 Welding and curing station
10 Amazon and king trailer
11 0.5 Tonne jib crane
12 60 Tonne crab crane
13 5 Tonne hoist
14 Mat hoist mechanism

FIGURE 3   WASTE PACKAGING BUILDING

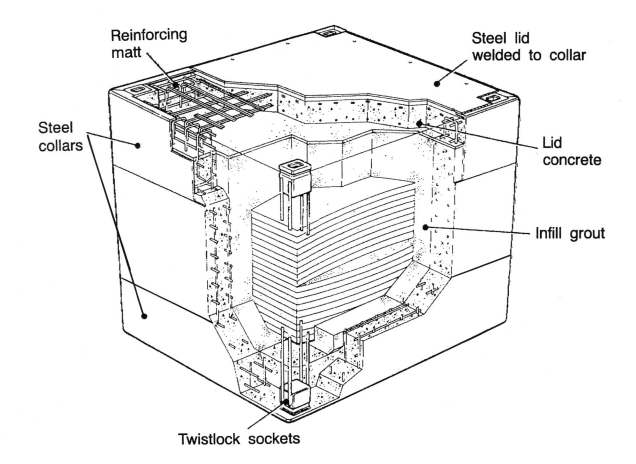

Reinforcing matt

Steel lid welded to collar

Steel collars

Lid concrete

Infill grout

Twistlock sockets

FIGURE 4 WAGR IL WASTE CONTAINER

C431/039

The interaction between the design of containers for low level radioactive waste and the design of a deep repository

S V BARLOW, CEng, MIMechE
United Kingdom Nirex Limited, Harwell, Didcot, Oxfordshire
P DONELAN, CEng, MIEI and T P DUTTON, CEng, MIMechE
Ove Arup and Partners, London
A F HUGHES, CEng, MIMechE
Risk Management Consultants Limited

SYNOPSIS United Kingdom Nirex Ltd is developing specifications for standard containers for the packaging of low and intermediate level radioactive waste for disposal in a deep underground repository. This paper describes the work undertaken on the interaction of the design of boxes for low level waste with the design parameters for the repository, in particular those for the shaft and disposal caverns.

1. INTRODUCTION

United Kingdom Nirex Limited (Nirex) has been established by the nuclear industry with the agreement of the Government, with the task of developing and operating a Deep Repository for disposal of solid low level and intermediate level radioactive wastes. Most intermediate level waste (ILW) will be packaged for deep disposal into 500 litre drums which will be transported to the repository in shielded containers designed to the requirements for IAEA Type B packages (1). Low level waste (LLW) and some intermediate level waste arising from the decommissioning of nuclear reactors will be transported in standard boxes designed to meet the requirements for Industrial Packages (1). These boxes serve not only as the transport package but also as the disposal container in the Deep Repository, and must therefore be designed to meet the requirements for both transport and disposal. In a recent paper (2) a description was given of the considerations which led to the choice of dimensions, weight and handling features of these boxes. This paper describes the work undertaken by Nirex to examine the interaction between design features of LLW boxes and the repository design in such areas as box size and weight, voidage and box stacking characteristics.

2. BOX SIZES

2.1 Constraints on Dimensions and Weight

2.1.1 Transport Considerations

In order to establish the optimum dimensions and weight for low level waste boxes it was necessary to consider constraints arising from the nature of the waste itself, and from its handling, transport and disposal. The bulk density of the waste to be packaged is clearly a major factor and much effort has been devoted to identifying the densities of different waste streams, taking full account of conditioning processes such as supercompaction. The dimensions of particular items or arrays of items such as supercompacted drums were also examined to ensure that a high packing efficiency could be achieved.

Constraints on the box weight and dimensions depend on the mode of transport adopted for the journey from the waste producer's site to the repository. Some sites have no rail access directly or have only limited sidings; it is therefore necessary to consider the requirements for both road and rail transport. Even if rail is the preferred transport mode, it may still be necessary to use road transport to marshall full-length trains at a convenient marshalling yard. For rail transport the box dimensions are limited by the W6 rail loading gauge while the weight depends on the axle weight limit; the preferred wagon would be a four-axle design with a payload of around 60te. The Construction and Use Regulations (3) govern the dimensions and weight limits for road transport; dimensional constraints are less onerous than for rail but payload is limited to around 25te with a 38te gross vehicle weight. Greater payloads may be carried using Special Type Vehicles (4) although these suffer constraints on crew size, speed, routing and prior notification of journeys.

2.1.2 Repository Design

Of particular interest in the present work were the limitations on box size arising from handling and disposal operations in the Nirex Deep Repository. The shaft providing access from the surface receipt facility to the underground caverns is significant in imposing limits on both box plan dimensions and weight.

A parametric study was carried out, based on present knowledge of geology at the candidate repository sites, to examine the relationship between box dimensions and the technology required for construction of a shaft to a depth of 1000m through water bearing rock. The study concluded that it was feasible to construct a shaft to accommodate a box of plan dimensions 4m x 2.4m but it was noted that this would require a moderate extension of current technology. A box of dimensions 5.5m x 2.4m would require a more significant extension of technology and hence entail greater development risks. The construction of larger shafts still, was considered to be very optimistic in that their construction although theoretically possible was way beyond present practice. In view of the development risks highlighted by this study the deep repository cannot plan at present on being able to accommodate a 6m long package (such as the half-height ISO container currently used for LLW disposal at the Drigg site in Cumbria), and a box of smaller dimensions will need to be specified.

The parametric study also examined the relationship between package weight and shaft hoisting technology and a payload limit of 50te was recommended. This was seen as the optimum package weight from a consideration of development risk. A 50te payload would still require an extension of technology but it was seen as a feasible rather than an optimistic design basis.

Shaft design is a function of depth, diameter and ground conditions and if for instance the depth of shaft were to be reduced or the ground conditions were better than postulated, then this would result in a reduction of the development risk. As detailed geological information is made available from the site investigation work it may be possible to re-evaluate the development risks and revise the limits on package length and weight upwards, but until such time, the maximum box length is taken to be 5.5m and the maximum allowable package weight is taken to be 50te.

The shaft is also a potential bottleneck for throughput of waste, so it is important to minimise the number of boxes to be lowered and to avoid a large number of different sized boxes with differing handling requirements. Advantages were identified in having a limited family of standard boxes sized to allow different modules from the family to be handled and stored together if required.

2.2 **Optimisation**

Having established the constraints on box sizes, a methodology was devised to identify optimum dimensions and weight in order to minimise the overall costs of packaging, transport and disposal. This methodology was used to examine the variation of overall cost with box laden weight. For most operations economies of scale apply suggesting that boxes should be made as large as possible. However, if the option of transport by standard road vehicles is to be retained, the gross laden weight should not exceed a figure in the region of 25 to 26te.

Knowing the bulk density values of the different waste streams and with an estimate of the likely box tare weight, it is possible to evaluate a set of payload volumes for a family of boxes. An examination of the density data indicated two distinct average values; however, in reality the bulk density in any given box will vary about this average. If density is higher than average the box would have to be underfilled to remain within weight limits, leading to transport and disposal of unfilled volume; on the other hand, if density is below average the box will not attain the optimum weight determined for transport and disposal operations. The methodology was used to examine the relationship between these two cost penalties for a range of distributions of bulk density. It was established that overall costs are minimised by adopting a box payload volume that gives an average laden box weight approximately 4te less than the 26te maximum for road transport. In this way a compromise is achieved in which the number of boxes that have to be underfilled is acceptably low while the majority of boxes are reasonably close to the target to allow efficient transport and disposal.

2.3 **Standard Box Sizes**

By an examination of the interacting constraints and cost implications of all operations undergone by waste boxes it was possible to identify optimum box weight and dimensions. For low level waste a modular family of two boxes was identified as optimum, the larger having nominal dimensions (l x w x h) 4.0 x 2.4 x 2.2m, the smaller being a half-length version. For no extra cost, it was identified that adopting precise plan dimensions of 4.013 x 2.438m and 1.969 x 2.438m would maintain a relationship with existing ISO freight containers (which are based on imperial dimensions) giving possible benefits in manufacture, handling and transport, especially as the ISO twistlock handling system is to be adopted for waste boxes. Furthermore, it may be feasible in the future to introduce a 6m long standard ISO freight container if the constraints on repository shaft construction can be relaxed.

3. VOIDAGE

For disposal of intermediate level waste it will be necessary to minimise macro voidage and maintain a highly alkaline environment within each waste package. This is necessary to limit the solubility of longer-lived radioactive isotopes in the waste as well as to maintain low corrosion rates for steel containers. These requirements can both be assisted by means of cementitious grout infill which will be provided in addition to vault backfill grout.

The radioactivity content of low level waste is small in comparison to intermediate level waste and is relatively insignificant in the post-closure radiological safety case for the Deep Repository. Although the advantages of grouting would be similar in nature to those for ILW, it is not possible to demonstrate any quantifiable benefit. To be weighed against any benefit is the fact that grouting would significantly increase the weight of a LLW box by up to a factor of two and will have sizeable cost penalties. The cost of operations to place the grout in the box must be considered together with implications for manufacture of any features in the box to facilitate grouting, and any strengthening required due to the increased weight. If grouting were to be carried out it would most easily be done by the waste producer when placing the waste in the box. However, this has been shown to more than double the transport cost, a significant element in the total cost. To avoid this, grouting could be done at the repository although this would require carrying out an operation on a box filled with waste elsewhere, dividing the responsibility for the package and complicating the quality control. Finally, the increased box weight would have cost implications for handling in the repository, particularly in shaft lowering operations. As noted above, the shaft represents a critical element in repository operations and grouting of LLW boxes could severely impede throughput; for instance two ungrouted 2m boxes could be handled in one operation compared with a single grouted box.

Another reason for void filling low level waste boxes with cement grout would be to assist in stacking of boxes in repository caverns and this is discussed in the next section.

4. STACKING OF WASTE BOXES IN THE REPOSITORY

4.1 Background

Initial conceptual designs for a deep repository located in hard rock, indicated that the construction of large caverns would be an economic option (5). The concept design proposed that packages would be emplaced underground in stacks of up to 35 metres in height and would be handled by overhead travelling crane. All packages, whether for ILW or LLW, were assumed to be fully infill grouted allowing these considerable stack heights to be achieved with relatively low compressive stress in the grouted waste.

More recently, the effect of leaving an ullage above the grouted waste within the package has been examined. Even with minimum ullage and assuming the stack is precisely aligned, the stacking performance of LLW boxes was found to be questionable over such heights. Various options were proposed to avoid this problem, one of which was to ensure that all boxes were fully grouted to eliminate ullage. However, as the grouting requirement from long term safety of LLW was being re-examined it was decided that the question of grouting for stack strength and the whole issue of box stacking characteristics should also be re-examined. The study has examined both the technical and economic aspects of waste box and repository design in order to determine the optimum design solution.

4.2 Box Design Options

4.2.1 Grouted Box

Conceptual designs of boxes for transport and disposal of low level waste to a deep repository have in the past been independently prepared by Croft Associates and Gravatom Projects Ltd. The Croft Associates design produced for Nirex is illustrated in Figure 1 and the Gravatom design produced for CEGB (now Nuclear Electric) is shown in Figure 2. Both concepts are based on the assumption that the contents would be infill grouted and that therefore the solid contents would be able to carry the loads imposed by stacking in deep underground caverns. The box side walls are provided

with strong stiffening members to resist the hydrostatic load of wet grout.

When the designs were re-evaluated to examine their performance without infill grout, it was shown that even with the resultant reduction in stack load, the corner posts were not strong enough to carry the loads imposed by stacking in the deep cavern. Further analysis showed however that if use could be made of the wall stiffening members, then the load carrying capacity of the box would be adequate. This analysis led to the concept of the 'strong-walled' box.

4.2.2 Strong Walled Box

The use of a 'strong-walled' box to carry stacking loads although a novel approach, is attractive since it does not require the provision of infill grout as an aid to stacking. It was recognised at an early stage that problems might arise through the need to ensure 'line contact' between boxes and problems associated with manufacturing and positional tolerance could be foreseen; a design study was therefore initiated to examine the feasibility of this concept. This study proposed various solutions but the preferred design comprised a box with rectangular hollow sections spaced at one metre inervals along its walls. Line contact was achieved via top and bottom rails and stack loads transferred via the hollow section members. The preferred design is illustrated in Figure 3.

4.2.3 Six-high Stacking Box

The other method of stacking which again does not require infill grout, is to make use of corner stacking posts as utilised in conventional ISO freight containers. As mentioned earlier, analysis has shown that simple corner posts cannot carry the stacking loads that would be imposed through stacking in a deep cavern and therefore a reduced stack height has to be specified.

Boxes for low level waste transport and disposal are to be designed as Industrial Packages (IP-2) under the IAEA Transport Regulations (1) and will therefore satisfy the structural performance requirements specified therein. With respect to stacking, the regulations stipulate that an Industrial Package should be able to resist a compressive load of five times its own weight; in other words boxes should be stackable up to six high. This therefore represents the minimum stack strength of a box. The design of such a box is relatively straight-forward and is well within the current practice for freight containers. Load transfer is achieved via four corner stacking posts and box walls are stiffened simply by the use of profiled wall panels. Such a box design is illustrated in Figure 4.

4.3 Repository Stacking Options

As noted earlier, initial repository concepts assumed construction of deep caverns and use of infill grout within boxes to achieve the necessary stacking performance (Figure 5). Boxes were assumed to be handled by overhead travelling crane and it was proposed that backfill grout would be poured between boxes after the emplacement of each layer. A similar deep cavern design could be utilised by use of the strong-walled box. In this case stacking loads would be transferred through the box structure and infill grout could be omitted.

More recent repository design studies have considered other options for the stacking of boxes underground. The first method is simply to adopt smaller caverns described as 'tunnel vaults', which are designed to suit the minimum box stack height of six high (Figure 6). Alternatively, a novel use of the backfill material has been proposed in which it is used to form load transfer structures within the full-sized cavern (Figure 7). In this concept boxes are again handled by overhead crane and stacked in the full-height cavern, but stack heights are limited to five boxes high. Gaps left between adjacent stacks are filled with backfill grout to form structural walls running the length of the cavern. These walls are then used to support a thick backfill slab poured over the top of the fifth box. The load imposed by this slab when it is poured has to be carried by the boxes below and even if rated for six-high stacking, the stack of boxes has to be limited to just five high. The structure thus formed can then be used to support a further stack of boxes again limited to a maximum of five high.

4.4   Optimisation

In order to establish the optimum box stacking characteristics, the various options for box and repository design have been assessed. For each option overall costs have been estimated and technical feasibility compared in order to identify the optimum design solution.

Three basic options can be identified:

Box Option 1:   Ungrouted 'strong-walled' box to stack in full-height caverns.

Box Option 2:   Grouted box to stack in full-height caverns.

Box Option 3:   Ungrouted box to stack 6 high in either tunnel vaults or full-height caverns with transfer structures.

A further option can be identified where grouted boxes containing an ullage space are stacked in full-height caverns. This option would however, require the use of a strong-walled box but with the added weight (and cost) of infill grout and therefore is not considered further.

An initial assessment of overall cost including box manufacture, transport and disposal, indicates that Box Option 1 is the most expensive owing to the cost of the box itself and its reduced payload. Furthermore, the problems of stack alignment and stability and the accumulation of tolerances over such a high stack would require further consideration.

Use of Box Option 2 depends on assurance that the grouted box contents can sustain the imposed stack load without any deformation or settlement which might interfere with stacking operations in the cavern crown. The cost difference between Box Options 2 and 3 is not great. Box Option 2 can take advantage of an economical cavern design and has the added benefit of not requiring accurate emplacement within the repository cavern. However, additional costs, both capital and operating are incurred by this option due to the need for the provision of a grouting plant. Box Option 3 on the other hand can utilise the cheapest of box designs and for the tunnel vault can make use of fork trucks for placement, although a disadvantage is that a larger underground footprint area is required. The use of full-height caverns with transfer structures is slightly more expensive due to the additional backfill material that is necessary and due to the additional cavern space that needs to be excavated. In addition, further work would be required to confirm the feasibility of construction of such structures using backfill material.

4.5   Preferred Stacking Option

Cost comparisons remain preliminary as many items are as yet poorly defined so it is important that the preferred option is not sensitive to these uncertainties. Of the options considered, Box Option 3 offers a feasible solution without requiring further development work. The box design is conventional and offers the capability of stacking up to six boxes high, a practice already well proven in the freight container industry and one which also meets the requirements of the IAEA Transport Regulations. Similarly lifting and handling can be achieved by use of well established technologies; overhead crane, fork truck and stacker truck handling are all possible and able to achieve the necessary accuracy of emplacement. The box can be transferred directly from the receipt building to the cavern without the need for operators to perform any intermediate grouting operations. The box itself is the cheapest of the three and as noted above, is credited with the lowest overall cost. In fact this box offers flexibility to repository designers as it may either be stacked in tunnel vaults or full-height caverns with transfer structures or floors. It is therefore the current intention that Box Option 3, a box to stack up to six high without the need for infill grout, be adopted as the basis for the Nirex standard LLW box.

5.   CONCLUSIONS

An examination of the technical constraints and costs associated with waste packaging, transport and disposal has allowed the optimum sizes for LLW boxes to be identified. A modular family of two boxes has been identified with nominal dimensions 4.0 x 2.4 x 2.2m and 2.0 x 2.4 x 2.2m. To maintain a relationship with existing ISO freight containers and thus to gain potential benefit in manufacture, handling and transport, precise plan dimensions of 4.013 x 2.438m and 1.969 x 2.438m have been specified.

It has been established that void filling of LLW boxes by cement grout or any other material is not essential for deep disposal. While grouting offers some mechanical advantages, any benefit must be weighed against the cost of grouting operations, the increased weight of the package in handling and disposal, and the effect on throughput and hence on overall disposal costs.

A study of the interaction between the design of LLW boxes and the design of the deep repository in terms of both cost and technical feasibility has indicated that the preferred option is to adopt boxes capable of being stacked up to six high without the need for infill grout. This solution offers flexibility to repository designers and the various options for tunnel vaults and deep caverns are being given further detailed consideration.

6. REFERENCES

(1)   INTERNATIONAL ATOMIC ENERGY AGENCY. Regulations for the safe transport of radioactive material, 1985 Edition. (As amended 1990). Safety Series No. 6.

(2)   BARLOW S.V., DONELAN P., DUTTON T.P., SMITH M.J.S. Boxes for the transport and disposal of low level and decommissioning intermediate level radioactive wastes. Proceedings of INE Conference, Transportation for the Nuclear Industry, May 1991.

(3)   HM GOVERNMENT. The Road Vehicles (Construction and Use) Regulations, 1986.

(4)   HM GOVERNMENT. The Motor Vehicles (Authorisation of Special Types) General Order, 1979.

(5)   TUFTON E.P.S. Design of underground works for deep repositories. Proceedings of BNES Conference, Radioactive Waste Management 2, Brighton, 2, pp 105 - 110, 1989.

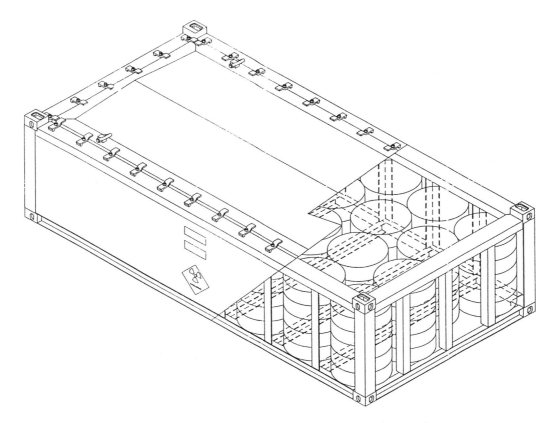

Figure 1. Conceptual Design of a LLW Box - Croft Associates

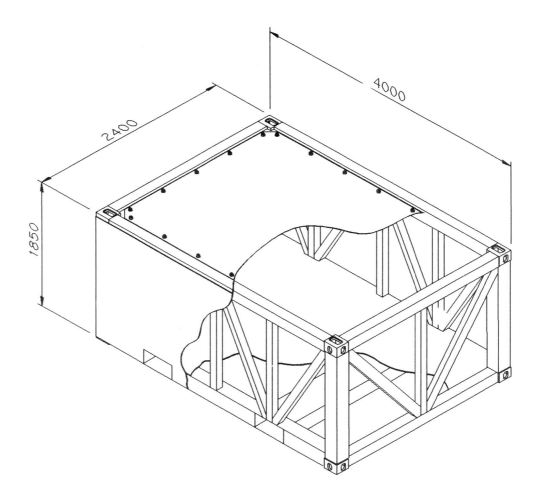

Figure 2. Conceptual Design of a LLW Box - Gravatom Projects

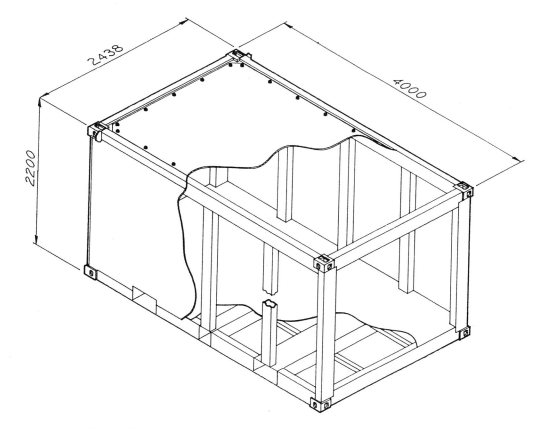

Figure 3. Conceptual Design of a Strong-Walled Box

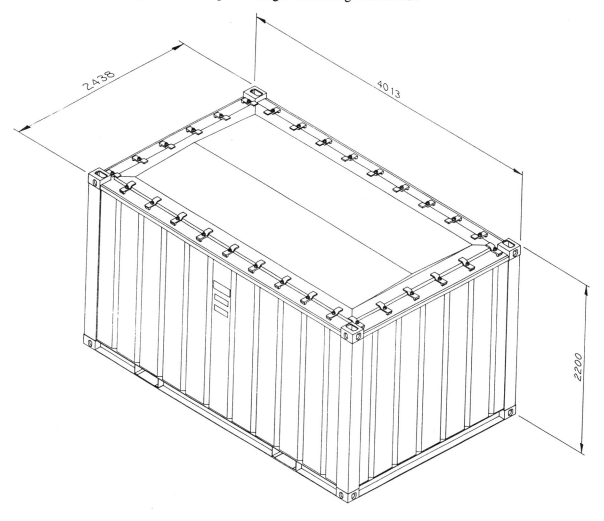

Figure 4. Conceptual Design of a LLW Box to Stack 6 High

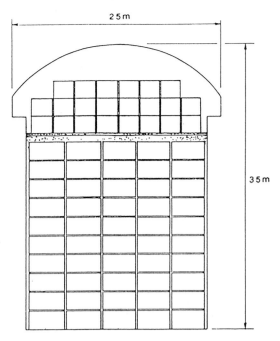

Figure 5. Possible Layout of a Deep Cavern

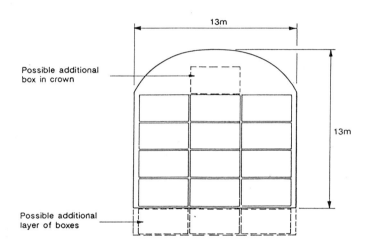

Figure 6. Possible Layout of a Tunnel Vault

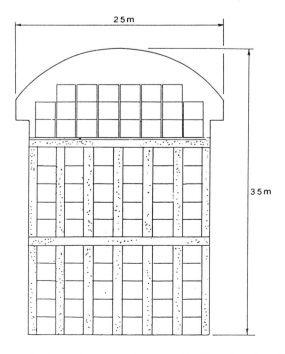

Figure 7. Concept Design of a Deep Cavern with Transfer Structure

C431/050

Use of robotics in a Radwaste treatment plant

C W E LEEKS, BSc, MSc, AMIMechE
AEA Engineering, Winfrith Technology Centre, Dorchester, Dorset

SYNOPSIS: A 762 Unimate Puma, clean room standard Robot has been installed and commissioned in the Radwaste Treatment Plant at the Winfrith Technology Centre. The robot interacts with a variety of purpose designed tools and proprietary welding equipment. It performs 13 dedicated tasks in the final closure and health physics operations, before the 500 litre waste drum is despatched from the plant.

1   INTRODUCTION

1.1 The Atomic Energy Authority* (AEA Engineering) Winfrith, Dorset have recently completed an on-site facility to immobilise intermediate level beta/gamma radioactive waste. The waste is immobilised in a cementation material within a 500 litre drum, before eventual disposal to a National repository. The drum and internal matrix have to conform to a national specification including dimensional envelope and a weight limit of 2 Te.

The Radwaste Treatment Plant (RTP) is approximately 25m long by 9m high and 9m wide, a cross section of the plant is included as Figure 1. The main features of the RTP will be highlighted in the following description, the robotic arm, the subject of this paper, is at the west end in the Drum Processing Area (DPA).

1.2 RTP Description

The RTP is arranged on three floors, the basement, rail level and cell level.

These floors are contained within the biological shield which is constructed from pre-cast reinforced portland concrete blocks, 1.2m thick and weighing up to 14 Te each. The unique module interlocking blocks are designed for an outer surface radiation level of 2.5 uSv/h.

The rail level, which traverses the plant at the building floor level, contains a mono track, on which runs trolleys carrying the 500 l waste drums. Along this track are 15 stations positioned throughout the length of the plant.

The drum trolleys are 1.4m square and designed for a payload of 15 Te. The trolley's movement is remotely controlled from the Plant's control room by the plant's Programmable Logic Controller (PLC).

In the control room the operator has no direct vision of the plant, but observes the plant through three monitor screens: they display graphic representation of the state of the plant. Two further close circuit television screens allow vision of the interior of the RTP.

The Plant Control System is based on a PLC made by General Electric (USA). It is a computer with a restricted function set designed specifically for process control applications.

Communication to the individual trolley is via infra red transmitters and receivers plus reflective position marker which are set between the rails. Positioned on each trolley, to align with the positional markers, are two sensors and one transmitter, together with a retro-reflective transmitter and receiver which acts in conjunction with the reflective marker to stop the trolley.

The Sort Cell positioned towards the East end of the RTP is the module where the solid waste is sorted and sentenced. This containment cell is 6m long by 4m wide by 4m high and constructed of 12mm thick plate. Four viewing windows, each with two master slave manipulators are used to sort the waste. The waste enters and leaves the cell through double lidded posting ports.

The cementitious material is added to the drum containing the sorted waste through the fill head.

The trolleys are parked for 24 hours to allow the matrix to continue curing.

---

* Formally trading as United Kingdom Atomic Energy Authority (UKAEA)

The final drum processing operations are carried out in the DPA and will be described in the next section.

2 BACKGROUND

Conceptual work for the Winfrith RTP has been undertaken since 1978. This has resulted in various studies into specific aspects of matrix compatibility, long term corrosion, radionuclide migration and similar work.

In 1983 conceptual designs for the RTP were started within Design Branch, Winfrith.

Although the finished plant is generally similar to this early design, one area has changed significantly and this is the DPA.

The initial design work had three, specifically designed multiple purpose 'gantry' devices incorporating mechanical, electrical, hydraulic and pneumatic systems. The devices straddled the mono track in series, and each performed various sequential tasks. The drum was progressively removed and returned to the trolley as each device remotely performed its tasks.

The processed drum eventually left the third device to be presented on a turntable for checking by health physicists for surface contamination. The suitably protected operators, behind a shield wall and window, would smear the drum. The smears then required checking.

As a result of a combination of the 1985 Ionizing Radiation Regulations, consideration of HAZOP studies and the individual detailed difficulties of commissioning three gantry/operational devices, the design for the drum processing area was reviewed.

This is a dilemma often faced by designers, when, during the design process does one freeze the conceptual phase; since new and emerging technologies are advancing at such a pace that they can make 'traditional' design immediately outdated.

The main process functions to be undertaken in the DPA are:-

(a) checking the curing matrix,
(b) welding the drum and lid,
(c) filling the void under the cap with grout,
(d) welding the final sealing cap to the lid,
(e) health physics smearing of the drum.

Design Branch, Winfrith studied the options available for the DPA.

One alternative was to replace one of the bioshield blocks with a shielded window and master slave manipulator. These could be used in conjunction with simplified dedicated plant.

A second option considered was to position the drum on a turntable and use a robotic arm to perform the required functions.

The study concluded that this option was preferred.

3 DEVELOPMENT WORK

3.1 A development programme was prepared to enhance confidence in the use of robotic equipment by:-

(a) studying the robotic arm market,
(b) confirming the robotic arm's suitability to carry out some of the major process requirements,
(c) to establish a full size development cell,
(d) to demonstrate that the robotic arm could be remotely removed from the RTP,
(e) to recommend equipment and plant such that the detail design of the RTP was not delayed,
(f) to work within a quality assurance programme.

A development specification was written that if successful would achieve the programme above.

The employment of a Robot System to do many tasks in a remote location and the tooling required to adapt it, makes its application fairly unique. Therefore much of the development specification has evolved from the experience gained during this work.

The study of the market place and the operating envelope required (both spatial and payload), led to the ultimate hiring of a Unimate Puma 762 robotic arm, for trials in the Design Branch development laboratory at Winfrith.

The system consists of a control unit with a visual display terminal and keyboard connected to the 6 axis articulated robotic arm.

The arm is capable of carrying loads up to 20kg at its furthest joint at full acceleration. A pneumatically powered gripper is fixed to the end of the arm. It is modified to locate each tool accurately and in a particular orientation.

The arm was initially hired since one objective was to consider its use in the radiation background levels of the plant.

Although the AEA has subsequently developed a radiation hardened robot, with Unimation (the NEATER robot) which is claimed to withstand 100 M Rad; the study of the estimated background radiation level has indicated an acceptable life within the RTP for a 'standard' robotic arm.

The DPA plant design incorporating the robotic arm.

(a) Trolley station - to stop drum trolley in DPA.
(b) Remotely controlled gantry - to move drum from trolley to within robot operating envelope.
(c) Turn table - to present drum to robot.
(d) Robotic Arm and dedicated tooling.
(e) Welding equipment.

A trolley and control system had already been developed with a stopping accuracy and repeatability better than ± 0.1mm. This was used in the development cell, together with a turn table needed since the robotic arm would not easily span the 0.8mm diameter drum.

Another early decision was the type of welding equipment to be used. The fundamental choice was between the Metal Inert Gas (MIG) or Tungsten Inert Gas (TIG) systems. Both pulsed synergic MIG and pulsed arc TIG were studied in detail. The remote (behind 1.2m concrete without manual intervention) application was different to most remote welding applications. Synergic pulsed MIG appeared to be more suitable because of its greater tolerance capabilities, minimum of spatter during welding and minimum in-cell maintenance requirements.

This equipment was installed in the development cell and simple welding trials started.

3.2 Welding Development

The welding development took over a year from specifying the plant to completing the final software package.

During this time the turn table drive and controller were changed to optimise its speed, for welding and torque capability. A GE series 3 PLC is used to monitor and control the turntable. It is programmed with a number of sequences which cater for all robot requirements. These sequences once selected by the robot, are autonomous until completed.

These trials also demonstrated that a certain amount of run-out was experienced, due to the out of roundness of the drum and lid weld specimens. The weld prep could not be followed accurately enough to maintain a continuous unbroken weld, therefore a method of following the welding seam had to be considered. A market survey of seam tracking devices was undertaken. The use of laser, opto-electrical devices proved to be expensive and required line of sight from outside of the biological shield. This would have occupied valuable space and could not have coped with existing tacking positions.

A method of 'mapping' the weld profile was developed at Winfrith. During the mapping sequence of each drum, prior to welding, it is necessary to stop the table at each of 32 selected locations during 1 rotation. The improved table drive proved that this could be achieved consistently and the stopping and starting 'ramps' could be adjusted to maintain accuracy, without reducing the rotational progression speed.

The mapping method utilises the consumable wire protruding from the welding torch contact tip as a probing element and stores the appropriate torch positions in the robot controller memory.

When the sensing element touches the workpiece a short circuit is produced in a low voltage circuit, creating an input signal to the robot controller. This signal is used by a program running in the controller to identify the presence of a surface. The location of the probe, held in the robot gripper, is stored in memory as an array of joint angles.

Following the mapping, the welding sequence commences, with the torch moving to its corresponding array position. This is repeated until the desired number of tacks are completed. As the turntable passes again through the zero position the continuous weld commences. Each of the optimum array positions is selected as the weld progresses through the 32 segments. A predetermined delay at the end of the run allows a short overlap of 'run-out' weld.

3.3 Drum Smearing Development

The drum's outer surface has to have a nominal 100% smearing to test for any loose surface contamination.

Many combinations of smear, tool holder and the full dexterity of the robotic arm were tried, together with conventional health physics methods, before a realistic system to reproduce the human action was developed.

The final system uses six smears. The robotic arm applies a constant wiping force through a parallel action smear holder to reach all the necessary drum surface.

3.4 Development Conclusion

During the development of tools and routines the robot arm proved that it could carry the loads required of it and that its locations were consistently repeated within an acceptable accuracy of 0.2mm.

The initial stages of the robot arm development indicated that the arm could be used for other operations.

At the end of the hire period the confidence gained in its ability to perform the tasks remotely was conclusive.

The tasks identified and demonstrated, through this development programme, that the robotic arm could accomplish in the RTP were:

(a) Tack welding of Splashplate
(b) Checking grout condition
(c) Removal of excess bleed water
(d) Piercing of ventilation holes in Splashplate
(e) Placing of drum lid onto the drum
(f) Cutting of welding wire prior to each 'mapping' sequence
(g) 'Mapping' of the welding prep profile
(h) Welding of drum lid to drum
(i) Placing of closure cap into grout filling hole
(j) Seal welding of closure cap to drum lid
(k) HP smearing of the drum surface
(l) Disposing of used smears
(m) Remote retrieval and replacement of the robotic arm.

As a result of this development programme a 'clean room' standard 762 Unimate Puma was purchased for the RTP.

4 RTP INTERFACE

4.1 Robot Arm Controller and Communications

The Robot's Video Display Terminal (VDT) is to be installed in the RTP Control Room for the plant operators use.

The VDT will be enabled by the plant operator from the master membrane keyboard. Once enabled the VDT will present the operator with a control menu allowing sequential access to all robot process routines. However, their initialisation will be subject to plant interlocks being in a satisfactory condition.

The system of communication to be used, consists of hard wired signal lines connected between the robot controller and a suitable Plant PLC I/O Rack.

A number of hard wired lines are used to enable the robot to provide the Plant PLC with two pieces of information:

(a) The sequence selected by the operator
(b) The current robot status.

The remaining lines are used to provide communication in the opposite direction, ie PLC to Robot. They provide signals to enable the VDT and also issue "Proceed" or "Wait" instructions dependent on the state of the plant interlocks. Failure of any specified interlocks will result in the Robot sequence resetting, followed by a display of the interlock error on the Graphics System.

This method of communication is simple to implement and avoids the purchase of expensive proprietary communications software.

4.2 Welding Power Source and Equipment

The majority of the welding equipment is situated outside of the RTP biological shield wall on the South West corner.

The welding unit consists of:

(a) Mains stabiliser
(b) 500 amp power unit
(c) Microprocessor (TR19)
(d) Robot interface
(e) Wire feed unit
(f) Water cooler
(g) Gas bottle (housed on trolley)

These items are housed within a steel cabinet with the exception of items (e) (wire feed unit) and (g) (gas bottle).

The wire feed unit is to be mounted on the bioshield wall in line with a penetration which will carry the umbilical to the welding torch head within RTP. A further umbilical connects the wire feed welding process quality assurance system.

Initiation of the welding parameters and welding control is signalled by the robot controller.

4.3 Welding Quality Assurance Monitor

The proposed method of monitoring the welding parameters is by installing an 'Arc-Guard' system. Arc-Guard is a microprocessor based system for the comprehensive monitoring of weld quality in robotic and other 'synergic' and MIG welding applications. The system can identify the production of defective welds due to a variety of causes and maintain a record of the parameters associated with each weld which has been performed.

The parameters monitored are:

(a) Weld current
(b) Weld voltage
(c) Wire stick-out
(d) Arc stability
(e) Gas shield integrity

Preset limits can be entered into the units program to produce signals for an alarm indication, or to abort at different selectable stages of the welding process. An end of weld report can be generated and stored for QA purposes.

A serial data line is provided to enable the off-loading of the QA data at suitable periods during the process campaign. This information is relayed to the Radwaste Operations Computer.

5 ROBOTIC ARM TASKS IN THE RADWASTE TREATMENT PLANT

5.1 Task 1 - Tack Welding of Splashplate

The waste process drum arrives at the turntable with the splashplate placed in position. The securing of the splashplate has to be carried out before any tooling or instrumentation is passed through it to ensure that it is not displaced. The position of the splashplate rim allows it to be secured using the 'Synergic' MIG welding system. A number of small welding tacks are applied to fulfil this requirement.

5.2 Task 2 - Checking Grout Condition

A quality assurance requirement of the processed waste and cement matrix is to ensure that a particular hardness is achieved after a predetermined time. The figures recorded can be used to assess the curing progression of the mix.

The hardness assessment is carried out using a tool designed to register the top surface of the mix and to monitor the mechanical resistance using an inbuilt calibrated load cell.

This tool is stored in the tool rack and is connected to an indicating instrument with an electrical cable.

The robot arm collects the tool from the rack and moves it to a position directly above a grommeted penetration within the process drum splashplate. The tool is then directed vertically downwards through the grommet for a programmed distance which will bring the probe into contact with the surface of the mix.

This produces a status signal to the robot controller to indicate the acceptance of the infill material hardness within the process drum.

The analogue signal sent to the PLC is converted into a digital reading giving the direct hardness figure.

5.3 Task 3 - Removal of Excess Bleed Water

If any surface water is present, then after the hardness acceptance it has to be removed prior to the injection of the capping grout. The hardness probe has incorporated in its design a dewatering facility, when opened this valve exposes the probe surface to a vacuum circuit which will pull any adjacent bleed water through the valve to a drain tank.

The hardness/dewatering tool is then returned to the tool rack by the robotic arm where the tool is subjected to a washing cycle.

5.4 Task 4 - Piercing of Ventilation Holes in Splashplate

The splashplate which forms a physical barrier during the active filling of the process drum, prevents contamination of plant and outer surfaces of the drum and must not trap any air voids within the capping grout. To maximise the amount of capping grout allowed into the drum it is necessary to vent the splashplate evenly to release the maximum amount of air. In selected positions within the splashplate are thin foil discs for this purpose.

A punch tool parked in the tool rack is collected by the robot to complete this foil piercing routine before it is returned to the tool rack.

5.5 Task 5 - Placing of Drum Lid onto the Drum

The drum lid is brought into the Plant on the drum trolley. The lid is supported on its edge and is set at the leading end of the drum trolley.

Collection of the drum lid from the trolley is carried out using the modified gripper which is an integral part of the robot arm. The gripper jaws are fitted with extended fingers which locate and expand within the central hole of the drum lid boss. The drum lid is raised vertically from its support and orientated into a horizontal attitude before positioning and releasing into the process drum recess. The tolerancing and profile of the drum lid and rim offer a lead during entry.

5.6 Task 6 - Cutting of Welding Wire

Prior to each 'mapping' sequence it is necessary to 'zero' the feed wire stick out length of the welding torch. The simplest method to achieve this is to cut it to a known length. This method also ensures a repeatable end feature and 'bright' finish to the end of the wire. A proprietary cropping tool at the welding torch parking position will carry out this function. The robot controller will initiate the wire feed system to eject a short length of wire from the welding torch nozzle, which will be cut back to a set length by the cropping tool.

5.7 Task 7 - 'Mapping' of the Welding Prep Profile

The attachment of the drum lid to the drum is to be carried out using a 'Synergic' MIG welding process. This system uses an automatically fed consumable electrode which has to be presented to the weld preparation within acceptable tolerances for a satisfactory weld. Tolerances within the drum manufacture and positioning onto the turntable cannot guarantee these limits.

Therefore the 'Mapping' process recognises the weld profile position and calculates the optimum setting for the feed wire during welding.

Method: Following the precise cropping to length of the welding wire the isolation of the welding power supply, a 24 volt supply is passed through the wire. When the wire makes contact with an earthed surface (ie the drum weld profile) the signal received is related to an arm position and is memorised by the robot controller. An algorithm within the software calculates the next

sensing approach and from the results calculates the position that the robot arm has to move to, to maintain the correct welding attitude and stand-off. This principle is adopted for 3 weld profile contact points between the process drum and lid at 32 discrete positions around the circumference. Rotation and angular control of the turntable achieves each of the discrete positions.

5.8 Task 8 - Welding of Drum Lid to Drum

The method used for the lid attachment is performed by a Synergic MIG process. Once the weld profile has been mapped and accepted as being within set dimensional limits, a programme is run which complete the weld. This programme is run as a continuation of the 'Mapping' sequence and welding commences when the zero position of the turntable is passed. The number of positions mapped is divided into a smaller number of equal positions. When the turntable progresses the weld prep to coincide with the welding torch at each of these points, a tack weld is made to secure the lid. This routine prepares the lid for a continuous weld. The position of the torch during this rotation is repeated from the memorised locations during the 'Mapping' routine. After completing this sequence and passing the zero position, the welder is prompted to start a continuous weld and to repeat the same mapping locations. At each location the robot arm will align the welding torch for its optimum welding position from the data derived during mapping.

The programme also selects the progression rate of the weld by the selection of turntable speed.

5.9 Task 9 - Placing of Closure Cap into Grout Filling Hole

As soon as the capping grout operation is complete (this is the only drum processing operation in the DPA with which the robot arm is not involved): the closure cap placing sequence commences. The closure cap is brought into the DPA on the same drum trolley as the process drum. It is supported upon a pedestal for easy access by the robot arm. A 'Pick and Place' tool is selected from the tool rack by the robot arm and is manoeuvred over the closure cap. A downward motion of the arm engages the cap into the tool. The assembly is transferred to the grout filling hole in the drum lid where it is located and placed by ejection from the tool.

5.10 Task 10 - Seal Welding of Closure Cap to Drum Lid

The closure cap sits directly onto the surface of the lid boss and is located by a central spigot which protrudes into the bore of the boss. The profile at the edge of the cap is suitable for a small fillet weld.

To identify the existence and the position of the cap prior to welding, the welding wire is cropped to length and a single 'Mapping' routine is carried out at the welding start position. Upon satisfactory recognition and positional acceptance of the cap the turntable is started from the robot controller and after one revolution another signal initiates and maintains the weld until it is complete. Upon completion of the weld the welding torch is returned to its parking position by the robot arm.

5.11 Task 11 - HP Smearing of the Drum Surface

Prior to leaving the RTP sealed drums must be checked for loose radioactive contamination. This is defined as less than 4 Bq/cm^2 - Beta Gamma.

Eight smears in a tray are brought into the DPA on a drum trolley at the same time as the shielded vessel and transferred to a working position using the robot arm. A smear holding tool parked in the tool rack is collected by the robot arm and is used to select a smear from the tray.

The method used to smear the drum vertical and top surfaces is with the drum rotating upon the turntable. A smear is pressed against the drum surface and is moved to produce a series of overlapping concentric paths over the lid. Six smears are used to smear nominally 100% of the drum surface. Each smear is to be monitored upon completion of its section. A BP4 Beta Scintillation counter housed in a shielded containment is to be used for this purpose.

A smear is entered into a carriage at the base of the monitor containment. The carriage and smear is raised into the underside of the containment to a monitoring position. It is held in this position for a scintillation count period (1 to 5 minutes).

During this time counting signals are relayed to the Control Room Graphics Display System:

(a) Analogue signal showing current measurements in counts per second

(b) Digital trip signal informing of unacceptable high count level.

After completion of the monitoring period the smear is exposed by lowering of the castle door and is collected by the robot for disposal into the smear bin.

5.12 Task 12 - Disposing of Used Smears

Each smear will be used once and will then be discharged for disposal.

The robot arm with smear holder collects the smear from the monitor carriage and places it in a bin attached below the smear tray. This tray will be transferred by the robot arm to the drum trolley for removal from the plant, once all six smears are collected.

5.13 Task 13 - Remote Retrieval and Replacement of the Robot Arm

As described, it can be seen that the Unimate Puma 762 robotic arm is a major tool in the final stages of the process drums. In the event of a malfunction or routine maintenance of the cell unit, and considering ALARA principles, it may be necessary to remove it from RTP to a radiation free area for repair. The decision for removal will be assessed from the presence of the background radiation in the area around the robot arm.

The robot arm is permanently mounted on a 4 wheeled trolley which is secured to a raised plinth. The height of the plinth allows optimum access of the arm to each of its routines and aligns its trolley with the top surface of a drum trolley.

For retrieval and replacement of the arm the drum trolley top place is replaced with the 'Robot Retrieval Adaptor Plate'. The drum trolley is then sent into the RTP to align with the robot station.

The robot arm is checked using the CCTV to ensure that the position of the arm is not in a hazardous position for removal and to confirm that any tooling with cables are not in the gripper. At this stage the removal sequence is initiated:

(a) Pneumatic cylinder on plinth opens latch on adapter plate
(b) Two hydraulic cylinders on plinth raise the robot arm clamping cylinders to release its trolley
(c) Central hydraulic cylinder on plinth pushes robot arm towards the drum trolley

(d) With robot arm in position against stops on drum trolley the pneumatic cylinder on the plinth is de-energised which disengages the transfer ram and locks the robot trolley onto the adapter plate

(e) The transfer ram cylinder completely retracts the ram into a safe position.

The drum trolley with the robot arm can then be motored out from the RTP.

Replacing of the robot arm to its working position is a reverse process to the one described above.

Note: The maintenance periods recommended are:

(a) Routine servicing every 1000 hours
(b) Complete overhaul every 5000 hours

The items of plant described in this paper have been installed and commissioned in the RTP.

During 1991 an inactive operating campaign will be completed.

6   DISCUSSION

The engineer faces a dilemma at the conceptual stage of any large new project, whether to use existing methods or to review emerging technologies.

This can be a difficult choice, particularly when the environment is remote and radioactive, as in this case. Thus technology reviews and planned development activities should be programmed into a project: a fall back proposal should always be a real option. Provided this is managed as the project progresses from conceptual to detail design then emerging technologies should be considered in many cases.

Applying good engineering and common sense with the usefulness of a robot has certainly resolved difficult radioactive waste handling.

© UKAEA 1991

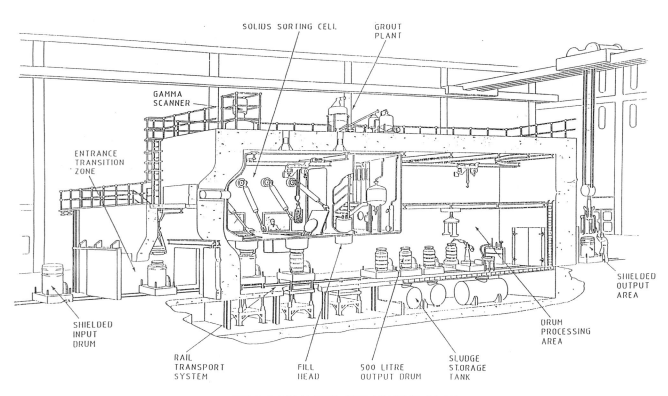

FIG.1 RADWASTE TREATMENT PLANT

C431/051

The crane handling system for 500 litre drums of cemented radioactive waste

A T STAPLES, BSc, CChem, MRSC, AMSaRS
Winfrith Technology Centre, Dorchester, Dorset

SYNOPSIS

As part of the AEA Technology strategy for dealing with radioactive wastes new waste treatment facilities are being built at the Winfrith Technology Centre (WTC), Dorset.

One of the facilities at WTC is the Treated Radwaste Store (TRS) which is designed to store sealed 500 litre capacity drums of treated waste for an interim period until the national disposal facility is operational.

Within the TRS two cranes have been incorporated, one spanning the entire width and travelling the length of the Store. The second operates within the area designated for drum handling during inspection work.

The development of the design of these cranes and their associated control systems, to meet the complex requirements of operations whilst also satisfying the reliability and safety criteria, is discussed within the paper.

INTRODUCTION

The Winfrith Technology Centre (WTC) in Dorset has accumulated a quantity of intermediate level radioactive waste over its operational lifetime. New facilities at WTC are currently being brought on line to encapsulate and store these wastes.

A cementation facility is undergoing commissioning trials at WTC. The Treated Radwaste Store (TRS), designed to store the 500 litre drums until a national repository is operational, has just been handed over to the WTC from the main building contractor.

The bulk of the waste consists of sludges formed during circuit decontamination and primary circuit water clean up of the steam generating heavy water moderated reactor (SGHWR). These sludges will be metered into a 500 litre stainless steel drum and mixed (lost paddle technique) with cementation powders prior to the drum lid being welded closed. The design intent was that a total of ~1000 of these drums will be produced in a three year period. A second waste stream consists of miscellaneous solid wastes which would be encapsulated directly in a similar 500 litre drum to the sludge drum. All these wastes are categorised lower level intermediate radioactive wastes; the approximate contact dose rates on the encapsulated waste drums is 20 mSv hr^{-1}.

1 THE TREATED RADWASTE STORE

The building, shown as fig 1 is approximately 20m high, 24m wide and 54m long and consists of a main hall formed by a clear span portal frame flanked by a single annex containing the boiler room, switch room and entrance. Internally the building consists of two areas, the main drum store and the core area consisting of receipt area, quality assessment (QA) facility and buffer store together with control room, change rooms and various plant rooms.

The superstructure generally comprises a metal clad structural steel portal frame. Internal steel framework for the core area is structurally independent of the portal frames; suspended floors in the core area comprises precast concrete units supported on a structural steel frame. Lateral forces imposed by wind, dead and live loads are resisted by the pinned base portal frames whilst resistance to longitudinal forces is provided via diagonal bracing.

The substructure is a stiff reinforced concrete slab, generally 1.5m thick which extends 2m from the perimeter gridlines on which the mainframe storerooms are centered.

Within the building a 5/15 te electric overhead travelling crane is provided, spanning the store and capable of travelling the entire length of the building.

The major functional areas of the building, the unloading bay, QA facility and main store are described in more detail below. A second 2½te crane operates within the QA area.

1.1 The Unloading Bay

The function of the unloading bay is to provide an area for controlled input of drums of encapsulated (weighing up to 2te) waste into the TRS; the removal of the drum in its shielding overpack being effected using the main 15/5 te overhead travelling crane. Empty overpacks (weighing 11 te) will be removed from the overpack area, and loaded onto a road transporter in the unloading bay, using the crane for re-use.

Overhead roller doors are provided to north and south for the entry and exit of the vehicle transporting the encapsulated waste drum in its shielding overpack. The unloading bay is approximately 16.5m x 5m x 20m high. Across part of its east face the unloading bay is open to the void over the QA facility, the division between the areas being a shield wall 4m high.

1.2 The QA Facility

This area receives the drum of encapsulated waste in the shielding overpack from the unloading bay. Here the drum is removed automatically and remotely from the overpack using the 5/15 te crane. The drum may then be removed to the QA line adjacent to the operator control room, this line consists of a number of plant items aligned along this wall. A 2.5 te travelling crane mounted off this wall, above these plant items is used to transfer the drum between the various stages of assessment. The equipment provided consists of:

- a rotary smearing turntable, electrically powered, enabling the drum to be rotated whilst a smear for external radioactive contamination is taken using a manipulator.

- master slave manipulator, operated from the control room located adjacent to the shielding window aligned with the smearing table.

- a movable trolley on fixed rails, onto which the drum is loaded, moving parallel to the shield wall of this area to transfer the drum to the radioactivity monitoring position.

- a bulk γ monitor.

- drum washing machine, for removal of any surface contamination detected on a drum.

- a stationary set down table onto which the drum is placed for the attachment of an interspacer (sacrificial mild steel plate intended to minimise drum corrosion during storage).

- a lifting magnet, used in conjunction with the 2.5 te crane to lift an interspacer onto a drum.

- interspacer stack.

The area is located to the east of the unloading bay, segregated by bioshield walls in an area approximately 10m x 12m x 10m high. The shield walls to the west and south are 4m high, those to the east and north 15.4m.

Three lead glass windows are provided at ground floor level on the north face of the area, providing a view from the control room to where the quality assessment equipment is located.

After removal from the overpack and from the quality assessment line the drum can be transferred using the 5/15 te overhead crane to be stored temporarily in the buffer store or to the drum lift area where the drum is manoeuvred into a "shielding chimney" via a chicane in the shield wall. The drum is then raised to the main storage area.

1.3 Main Drum Store

The main drum store is located east of the quality assessment area and is contained by shield walls to the north, south and west. It is approximately 14.5m x 17.2m x 20m high. Within the shield walls there is an array of storage tubes, of height 12.9m and diameter 972mm and capable of containing a stack of 9 drums of encapsulated waste. A total of 165 storage tubes are provided, but a double row of tubes at the eastern end of the array are filled with granular shielding material capable of being removed should the store subsequently be extended.

The individual tubes are of a modular construction and are made from precast concrete tubular sections, grouted together. The tubes are capped with a hexagonal interlocking removable concrete plug (one per tube) 950mm thick. These are removed using the main overhead travelling crane when access is required to a tube.

2 STORE DESIGN FOR MECHANICAL HANDLING

A requirement specification for the mechanical handling systems was issued with the sub contractor Strachan & Henshaw responsible for detailed design work. It was also the responsibility of the sub-contractor to show the interlocking systems for the handling systems complied with that specified in a CEGB standard. All important design parameters have been subject to independent review.

There were two basic elements of mechanical handling system design criteria - economic aspects and safety aspects.

From an economic point of view there were a fixed number of unit operations that had to be performed within the TRS and the number of mechanical systems necessary enabling them to be carried out was optimised. This resulted in the definition of two crane systems, the QA line crane and the 5/15 te overhead crane. The lifting feature (the grapple) associated with the 5 te hoist and the QA line hoist is identical and inter-changable; all operations performed with these hoists are via identically designed lifting features (shield plugs; waste drum tops; overpack lid tops, magnet). The overpack has a lifting feature which couples with the 15 te hoist bottom block assembly.

The TRS was designed against various safety criteria mainly derived from the NII safety assessment principles for nuclear chemical plant. It was clear that the mechanical handling systems would contribute significantly to the overall risk associated with the Store. This in turn has led to the incorporation of various design features in the handling systems to minimise the overall fault frequencies associated with the handling system.

2.1 The 5/15 te Overhead Travelling Crane

The crane spans the entire width of the Store and travels along its full length. Two hoists are provided on a single crab unit. For safety considerations a number of features are present which will both minimise service failure and possible hazardous events.

Both long and cross travel drives are similar in design. A main electrical incomer feeds a variable speed service motor which has a ramped controller for acceleration and deceleration (the control function defined by an experimental programme to optimise potential load swing). The output is fed via a chain coupling to the input of the drive gearbox, a second input to the gearbox passed via a drum brake of the energise to release type to the output of the fixed speed squirrel cage emergency motor which in turn is fed from a separate electrical incomer. The drive gearbox output powers the crane drive wheels. During operation both service and emergency motors rotate.

A non powered wheel for both long and cross travel is connected to an absolute position resolver the output of which provides positional information to the crane control plc. End of travel switches are provided to protect the crane in the event of plc position control failure. Further, six long travel switches and eight cross travel switches are used within the crane relay hardware to define certain zones within the Store; these zones in conjunction with directly switched height information from the hoist systems permit/prevent specific actions occurring. This is in addition to any software control.

The 15 te hoist drive consists of a single electrical incomer driving the service motor, the output of which drives via an eddy current brake which is in closed loop control with a tachogenerator driven by the main motor output. Output from the eddy current brake feeds via the service drum brake to the input shaft of the main gearbox; further an input shaft extension connects to a standby drum brake. The gearbox output shaft couples to the rope drum the other side of which is connected to the emergency disc braking system. The 5 te hoist is to a large extent similar except that an emergency electric motor is provided from an independent supply driving via an electromagnetic clutch. The separate motors being provided for long and cross travel and 5te hoist were defined as required by failure rate analysis and consequent dose uptake during recovery. No credible dose uptake may be apportioned due to 15 te hoist failure.

The major hazards identified to this handling system particular to the TRS concern operation with an unshielded drum above the shielding wall and direct interaction with the 2½ te crane. Both of these particular potential hazard frequencies have been minimized by use of the crane zoning interlocks.

2.2 The 2½ te Crane

Mounted along the face of the shield wall between the QA area and the control room are the top and bottom tracks which are provided to guide and support the crane in its east, west travel. Mounted on the bottom track are proximity switches, limit switches and end stops. Running along these tracks is the travel carriage of the crane which is fabricated generally from mild steel plate and forms an inverted right triangle. East, west movement of the travel carriage is accomplished via an endless loop wire drive system. All drives for the crane are located outside the shield wall in the unloading bay area. The hoist drive unit consists of a single rope system which again is driven from within the unloading bay and terminated in the west end of the QA area attached to a load cell. Both drives are transmitted via a pulley system across the shield wall and into the QA area.

The carriage drive is provided via a single a.c. electric brake motor, with interlocked provision for handwind via a helical gear to a wind on-wind off rope drum. End of travel limits are provided by limit switches which provide normal end of travel signal to which the operators may directly respond; ultimate end of travel switches are also provided which are direct acting and reversal from such a position can only be achieved by management intervention. Between east and west limits a series of six position switches are located which identify positions for various QA operations. No position back up is provided for this travel as failure to position the payload correctly cannot conceivably cause a significant hazard.

The hoist drive system consists of an a. c. squirrel cage motor with interlocked handwind facility connected via the service break to the input side of the reduction gearbox, an extension of the gearbox input shaft connects to the standby brake, both brakes being electromagnetic energise to release type. The gearbox output driving the rope drum. A proximity switch counting the serrated edge of the rope drum provides a measure for hoist overspeed detection and an extension to the

drum shaft drives a ten switch proximity detector for height control. A direct acting overraise switch tripped by the grapple within the QA area provides protection from normal height switch failure causing a potentially serious hazard.

2.3 The Grapple

The grapple forms the interface between crane and object and has been designed using the principle of conservatism and is simple yet robust. The grapple is shown as fig 2, is a three jaw device which locates on the lifting feature of the load. The jaws themselves are suspended from an upper grapple assembly which transmits the payload from the jaws to the hoist ropes. The jaws are separated by a lower assembly, about which the jaws are able to pivot. The lower assembly is in two parts, the upper plate providing location for the jaws, and the lower plate which is free to move relative to the upper section is supported by the jaws. To prevent movement of the upper and lower plates of the lower assembly a locking plate is used. The plate is spring loaded and may be released by solenoid (dual solenoids for redundancy). With the upper and lower plates at maximum separation the locking plate cannot prevent relative movement of the plates. When the plates come into contact, however, the locking plate locates beneath the shoulder of the centre spindle of the lower plate and locks the lower and upper plates in this configuration. The locking plate can only be released by activation of the solenoid.

The grapple design utilises the self weight of the mechanism to cause it to close and open. The closing operation is due to the weight of the upper grapple assembly and hoist ropes causing the over assembly plates to move closer together and hence cause the jaws to close. The force required to lower the upper base plate with the lower base plate supported is less than the weight of the upper grapple assembly.

During the opening operation the jaws must move positively away from the drum as the grapple is raised with the locking plate in the release position. This operation relies upon the self weight of the grapple and the requirement that the lower base plate provides a greater opening moment on the jaws than the closing moment caused by the upper base plate.

Failure of the grapple resulting in release of the payload (excluding hoist failures) is dominated by mechanical failure of component parts. Since the grapple was designed using the principle of conservatism and as the number of moving parts has been minimised the failure rate is estimated at less than 10^{-6} failures per year. Failure due to spurious operation when supporting a payload is considered incredible because of the mechanical operation of the grapple; spurious operation whilst the load is supported by the ground will not pose a hazard. Failure to open the grapple on demand could cause a hazard with failure probability dominated by solenoid failure resulting in non release of the locking plate. Emergency grapple release areas are provided along the QA line and in the main drum store to mitigate this failure route.

In summary hazards arising from mechanical grapple failures are considered to be acceptably small whilst installed means are provided to mitigate against failure of the solenoids for grapple opening. Spurious operation of the grapple cannot credibly result in a hazard.

3 AFFECTS ON SAFETY

It is encumbant upon plant operators to provide evidence that proposed systems can be operated without posing undue risks to both the workforce and members of the public both in normal and accident situations.

The TRS has been assessed against detailed criteria, some of the more relevant ones to the subject matter of the paper are described below.

- The average effective dose to the workforce in normal and probable unplanned events (noted as those occurring at intervals less than once per ten years) should not exceed 5 mSv yr^{-1}.

  This has been achieved with doses in normal operation from the mechanical handling systems providing no significant dose uptake. Further, detailed fault tree analysis of possible failure modes has not identified any fault which occurs at intervals less than once per ten years which would give rise to any recovery dose uptake.

- The time averaged dose commitment to any member of the workforce arising from radiological accidents should not exceed 0.5 mSv per year.

  The dose total resultant from all assessed accidents expressed as a percentage of the criterion was about 5%. Of that total the dominant contributory accident type was that occurring during general maintenance. Accidents with the handling system contributed only 15% of the total.

- The time averaged dose commitment to the most exposed person off site should not exceed 0.1 mSv per year (from accidents occurring on the whole site); for the TRS the dose commitment should not exceed 0.001 mSv per year.

  Again this criterion was met but this time the dominant contributory accident type was with the handling system. This time the total dose resultant from all assessed accidents expressed as a percentage of the criterion was only about 1%.

4 CONCLUSION

A robust mechanical handling system for handling 500 litre drums of cemented waste has been installed in the TRS at Winfrith. The design of the system has been assessed and has been shown to have no significant affect on the overall safety assessment criteria for the TRS.

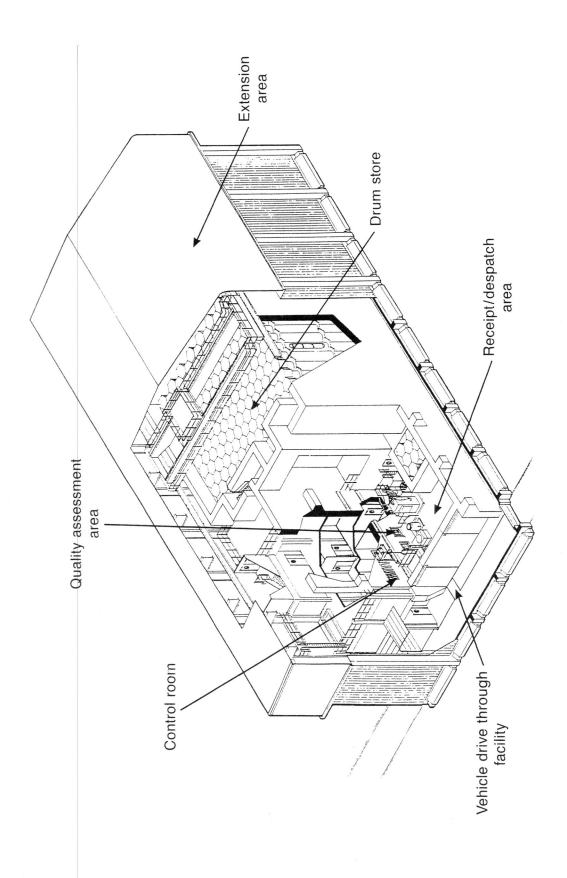

Extension area

Drum store

Quality assessment area

Control room

Receipt/despatch area

Vehicle drive through facility

FIG 1   TREATED RADWASTE STORE AEE WINFRITH

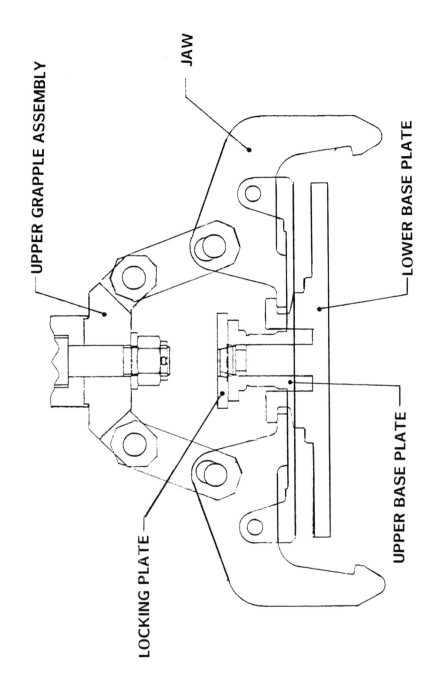

UPPER GRAPPLE ASSEMBLY

JAW

LOCKING PLATE

UPPER BASE PLATE

LOWER BASE PLATE

FIG 2 GRAPPLE (SHOWN FULLY OPEN)

C431/067

The design of a lifting grab for handling standard 500 litre radioactive waste drums

R PRATT, CEng, MIMechE, R MEDLOCK, BEng and D C STUBBS, BSc, MSc
Gravatom Projects, Fareham, Hampshire
R W T SIEVWRIGHT, BSc
UK Nirex Limited, Harwell, Didcot, Oxfordshire

SYNOPSIS

Within the present policy for future radioactive waste storage, conditioned intermediate level waste will be packaged into 500 litre drums. Gravatom Projects Limited have designed a standard grab capable of vertically lifting a range of 500 litre drums compatible with all stages of the proposed waste management route. This paper covers the assessment of various options through to the detail design of the grab.

1. INTRODUCTION

UK Nirex Ltd has been formed by the nuclear industry to develop and operate a deep repository for the safe disposal of both low and intermediate level radioactive wastes. The majority of conditioned intermediate level wastes will be packaged into standard 500 litre drums, and transported within approved shielded transport containers (Figure 1) from the various nuclear sites to the repository presently being planned.

On arrival at the repository these drums need to be lifted out of the transport containers and handled individually through the various stages of inspection before final disposal. Equipment is therefore required to carry out these lifting and handling operations, ideally in the form of a single lifting grab which can be used for all the 500 litre drums received.

The design parameters for these drums are set out in the Nirex 500 litre drum specification. This sets limits on the overall dimensions, weight, content and performance of the drum. In particular the specification allows for a limited range of basic drum shapes, each with a common design of lifting feature (Figure 2). This feature is provided in the form of a flange which may be from 750 to 800 mm outside diameter, up to 55 mm thick and situated up to 55 mm from the top of the drum.

UK Nirex Ltd commissioned a study with Gravatom Projects Ltd to develop a single design of drum lifting grab to handle all drums designed to comply with the Nirex 500 litre drum specification. This study was to identify a number of design solutions; determine the limitations, benefits, consequences and practicality of each design; and produce a set of manufacturing drawings and supporting calculations for a single selected design.

2. DESIGN CONSIDERATIONS

The work was undertaken in accordance with a specification generated by UK Nirex Ltd. Compliance with this specification was assured by the operation of a BS 5750 part 1 Quality Assurance System at Gravatom Projects, who have established a framework of procedures which ensure that the work is correctly controlled, documented and carried out.

The principal design considerations for the grab are as follows:
- 3t weight limit
- remote handling and maintenance
- radial dimension restrictions for grab
- automatic/remote operation
- minimum 3-point lift
- minimum radial force
- radiation tolerance
- material selection

These considerations are individually explained in more detail.

2.1 Weight Limit

Although a 2t limit is currently set for a 500 litre drum, the grab was designed to accommodate a 3t weight limit to allow for possible future uprating. In considering snatch loading, all components were designed for a 6t static load, allowing a factor of 2 for static to dynamic loading.

2.2 Remote Handling and Maintenance

Owing to the radiation levels associated with filled 500 litre drums remote handling facilities will be required for waste packaging activities. Therefore the design of the grab needs to be compatible with such conditions. Design points for consideration are as follows:-
- remote or 'gloved hand' maintenance, requiring ease of access to all components
- minimisation of the collection and trapping of contamination, and ease of decontamination
- ease of replacement of worn or damaged parts

2.3 Radial Restriction

To facilitate loading and unloading of 500 litre drums into a transport container, the open grab must operate within 40 mm radial clearance of the largest drum flange.

2.4 Auto/Remote Operation

The grab design includes features which are necessary for automatic or remote operation. The grab jaws are actuated electrically with sensors giving positive feedback of jaw position. Interlocks are provided to ensure safe operation of the grab. In case of grab failure, when attached to a drum, the grab can be disengaged and recovered by remote means.

2.5 Three Point Lift

The drum lifting flanges were designed on the basis of a minimum of three lifting points. For any multi-point lifting system it is debatable as to how loads will be distributed throughout the lift points as this depends on the flexibility of both the drum flange and the grab and their accuracy of manufacture. A three point system will be evenly loaded if the Centre of Gravity of the load is geometrically central. Increasing the number of lift points to four can increase the stress levels because all the load could be taken on two diametrically opposite points.

2.6 Radial Force

Conventional multi-point lifting systems suspended from a single point generate significant horizontal forces when a compact design is required. Operational constraints restrict the size of the contact members, which would result in unacceptable stress levels if exposed to high radial loads. Therefore it is necessary for the grab design to impose a minimal radial load on the drum.

2.7 Radiation Tolerance

Over the lifetime of the grab it may be exposed to a significant accumulated radiation dose. Therefore all of the materials of construction of the grab must be proven to be radiation tolerant.

2.8 Material Selection

The principal material of construction of the grab is austenitic stainless steel. This material is selected because it affords good corrosion resistance without the need for a surface coating that might give problems of decontamination and periodic weld inspection.

3. ALTERNATIVE DESIGN OPTIONS

Several alternative designs were considered in a preliminary study.
Whilst all were potentially acceptable, the requirement to accommodate a range of drum sizes and lifting flange dimensions, together with constraints on annular space allowed for the grab components round the drum, led either to high stress levels or to over complex designs. The five preliminary designs which were not adopted are briefly described and assessed.

3.1 Option 1

The lifting features in this design were vertical legs with an offset tongue. Engagement with the drum lifting flange was made by rotating the legs through 90°, such that the tongues are positioned under the flange. (Figure 3)

To accommodate the variations in drum diameter whilst staying within the 40 mm annular envelope, it would have been necessary to develop radial movement for the legs as well as rotation. This was seen judged to be unnecessarily complex and the design was abandoned.

3.2 Option 2

This design was based on a number of claws which slide radially to engage under the drum flange. Radial motion was provided by an electric motor actuator operating through a cam plate.
The action of drum lifting would maintain a radial load inwards onto the drum, through the inclined lifting arms, ensuring that claw engagement is maintained.

Problem areas with this design included high stresses in the claw assemblies, necessitating the use of at least 5 claws. Also, to optimise radial loads on the drum flange, a complex cam actuation system would have been required to balance the radial loads applied through the drum lift. For these reasons, this design was not developed.

3.3 Option 3

This design featured a claw arrangement within which a latch mechanism was housed. When the grab is being lowered, the latches would be restrained within the claw body. When the grab rests on the drum, the latches would be rotated outwards, pivoting on a pin, to engage on the underside of the flange. Careful design of the latch would have ensured that no loads are carried on the pin during drum lifting. (Figure 4)

This design met the annular space requirements. However, it was not possible to develop the design to accommodate the required range of drum sizes. Also, high stress levels in the latch and claw components would necessitate the use of either high-strength materials or more than 3 latches.

3.4 Option 4

This design featured claws retained and operated by a parallel link system, allowing the claws to operate vertically to engage on the drum flange. (Figure 5)

The grab would be self-centering when lowered onto the drum. Rotation of a cam plate by an electric actuator would allow the claw to fall under its own weight to engage on the drum flange. The claw design would maintain a self-sustaining gripping action driven by the weight of the drum being lifted.

High stresses were predicted in the claw components, and the lack of positive engagement onto the drum precluded further development of this design.

3.5 Option 5

The next design option employing the same parallel claw system as option 4 would allow for positive engagement of the claws onto the drum to be maintained during the drum lift. This was accomplished by linking the claws to the grab lifting feature, using inclined arms attached to the claws.

The weight of the grab lowered onto the drum would open the claws, and when lifted the weight of the drum would close the claws positively onto the drum flange.

Again, stress levels in the claw details were predicted to be high. The principal disadvantage, however, was the possibility that the drum could be released in the event of a collision.

4. GENERAL DESCRIPTION OF GRAB

The grab design that was selected for further work is illustrated in Figure 6. It consists of three main components, a fabricated body from which the claws extend, a fabricated bridge type structure from which the body is supported and the claw actuating mechanism.

The fabricated body locates the bearing housings, in which the claws slide, and the guides which centre the grab on the drum. These guides also protect the claws in the event of the grab colliding with an obstruction. Three feet mounted on the underside of the body support the grab when it is lowered onto a drum.

The grab claws are actuated by a cam and follower mechanism mounted within the body. Cam followers, mounted on the horizontal part of each claw, engage in tracks in a circular cam plate, which rotates on a shaft mounted through the centre of the body. The cam plate is driven by a motor and double worm gearbox mounted above the body.

The body is supported from a three armed bridge, fabricated from I-section beam, which spans the body and has a mounting lug for a bow shackle used for lifting.

The overall dimensions of the grab assembly are approximately 880 mm diameter x 750 mm high and the assembly weighs 230 kg.

5. OPERATION AND CONTROL OF GRAB

5.1 Sensors

A number of sensors are fitted to the grab. The prime function of the sensors is to provide feedback to the operator, but they could also be used as inputs to a crane and grab control system.
Proximity switches are used to indicate contact between the three feet of the grab body and the top of the drum. Further proximity switches are mounted on the 3 claws of the grab. Because of the allowable variations in flange thickness, 2 switches are required on each grab arm.

A proximity switch is also used to indicate whether the grab is fully opened. In this case the detector is mounted on the body of the grab, and is triggered by a feature on the cam plate shaft drive.

Finally a load cell is incorporated in the grab lifting shackle. This prevents opening of the grab when a drum is being lifted, by interlocking the load cell signal into the grab control system.

Although not illustrated in the diagram, it is proposed to route all wiring from the sensors through a common terminal box mounted on the grab body.

5.2 Motor Control

In order to ensure that the drum is securely held in the grab claws, it is proposed to control the grab motor using a current sensing device, which would switch off the power supplied to the grab motor at a pre-set value. By limiting the available torque from the motor, the radial load applied to the drum by the grab claws can also be limited. The current limit should be set to allow a maximum radial force of 100 kg to be applied, sufficient to hold the drum securely. Variations in frictional forces within claw mechanism during operation of the grab between service intervals are not expected to be significant.

5.3 Crane/Overhead Gantry Control

Whatever system is employed to support the grab whether a simple crane, telescopic mast or purpose built equipment one further control is required to ensure the safe operation of the grab.
Because of the allowable variations in drum flange thickness, from 15 to 55 mm, it is necessary to raise the grab at a low, or creep speed until contact is made between the claws of the grab and the underside of the flange. Contact between claws and flange will be indicated by a signal from the load cell. Once this is done, the grab and drum can be raised at full speed.

This procedure will eliminate snatch or shock loading to the grab and drum flange.

6. DETAILED DESIGN ASPECTS

6.1 Materials of Construction

Generally, all components are made from 316 stainless steel, with the exception of proprietary parts such as the motor and gearbox. Since the cam plate and drive shaft are relatively highly stressed components, heat treated martensite stainless steel has been chosen. This exhibits high strength and good impact properties down to -20°C.

Bearings for the claws are made from aluminium bronze, which has good corrosion and wear resistance. The jaw cam plate is driven by a 0.18 kW 3-phase motor driving through a close-coupled double worm gearbox with an overall reduction ratio of 600. With an output speed of 2.33 rpm, and a 90 degree rotation of the cam plate, the claws will engage the drum from fully open in about 6 seconds. The required output torque to apply a radial claw load to the drum of 100 kg is 90 Nm, compared to the 210 Nm available from the motor/gearbox. The high reduction ratio eliminates the need for a braked motor, as back running is not possible.

The gearbox is shaft mounted, and rotation is prevented by a torque arm attached to the body top plate. In the unlikely event of a failure, the gearbox and motor can be freed to allow rotation by simply removing a pin from one end of the torque arm fixing. Rotation of the complete gearbox assembly will then free the claws from the drum. These operations are entirely feasible using a remote handling device.

7. GRAB LOADING AND DESIGN STRESSES

The grab has been designed for a safe working load of 3 tonnes. A load factor of two has been assumed to allow for snatch lifting giving a design load of 6 tonnes. A complete set of design calculations has been produced showing that all resultant stresses at the design load are less than two-thirds of the material yield stress, or a factor of 5 less than the material ultimate tensile stress at the design load. The design also ensures that stresses in those parts subject to cyclic loading are below the fatigue limit for the material.

To assess the likely loads on the cam plate drive mechanism, in an accident condition, it has been assumed that a collision between the grab and an obstruction would result in a load on one claw of up to one tonne (this is equivalent to a 3 tonne mass being stopped in 0.1 seconds from 0.33 m/s). This results in a torque of approximately 316 Nm on the cam drive components. The cam plate drive components have been designed to cope safely with the full motor gearbox unit stall torque of 360 Nm therefore no problems are foreseen to arise from the assumed accident condition.

8. RADIATION TOLERANCE OF GRAB COMPONENTS

Both stainless steel and mild steel incur negligible damage even when subject to very high doses of gamma radiation. Therefore the general structure of the grab should be unaffected by radiation.

Because of the detrimental effects of radiation on normal lubricants, radiation-resistant oils and greases have been specified for the moving parts of the grab. These provide adequate lubrication qualities at accumulated doses up to 5 MGy -typical (5 x 10^8 rads).

The electrical insulation on the motor windings will also be susceptible to radiation damage. No information is available on the specific materials used in the selected motor, but generally the insulation in electric motors begins to deteriorate at integrated doses greater than 1 MGy (10^8 Rads).

The proximity sensors are of the intrinsically safe design which contains only the oscillators in the head, the amplifier sections being mounted remotely, and hence out of the radiation fields. Recorded data suggest that proximity sensors of this type function satisfactorily up to 10^4 Gy (10^6 Rads). Beyond this level some slight increase in sensing distance was noticed.

The proximity switches mounted on the grab feet and claws will probably be the most vulnerable to radiation damage; as they will be closest to the drum during operation. These sensors are mounted in polybutylene terephthalate casings. No radiation tolerance data is available for this material, although similar materials will withstand doses up to 1 MGy (10^8 Rads) with negligible damage.

The cables which are integral with the sensors are normally sheathed in PVC. Although relatively stable to radiation, up to 5 x 10^7 G_y (5 x 10^7 Rads) there is evidence to suggest that degradation of the PVC can lead to the formation of hydrogen chloride gas, with subsequent risk of corrosion and stress cracking of stainless steels. Therefore in high dose applications, consideration should be given to the use of alternative sheath materials.

The load cell incorporates a strain gauge, which is connected to a bridge circuit. The output from the bridge circuit is fed to a remote amplifier located away from the radiation source. The bridge circuit made up of resistors is expected to be stable up to 10^4 Gy (10^6 Rads).

9. MAINTENANCE

The grab has been designed for the minimum necessary maintenance and maximum life. All the bearings used on the grab have rated load capacities well above their normal operating loads. In the case of the claw bearings these are only loaded statically and therefore wear of these bearings should be minimal. The only other wearing surfaces will be where the claws contact the drum and the slots in the cam plate, but in normal operation these areas are relatively lightly loaded. Similarly the gearbox would normally operate well below its rated load. The electric motor, which is expected to be the item most likely to need replacement from time to time, is easily accessible; it should be possible to replace this item quite quickly, even using gloved hands.

It is not possible to give a reliable estimate of the likely life of components, since this will depend to a large extent on the actual duty cycle and the levels of radiation to which the grab is exposed. Similarly the maintenance schedule is likely to be determined by the levels of radiation to which the grab is exposed. However, it should be possible to operate the grab up to an integrated dose of 10^4 Gy (10^6 Rads) before any significant radiation damage is encountered.

10. CONCLUSION

After consideration of a number of options, the chosen design best satisfies the requirements for a drum lifting grab as specified by UK Nirex Ltd. Compliance with this specification was assured by Gravatom Projects Ltd's in-house Q.A. design procedures.

It has been demonstrated that a single design of drum lifting grab can be produced for handling all standard 500 litre radioactive waste drums.

Sealed Transport Container Concepts for Drummed Intermediate Level Wastes

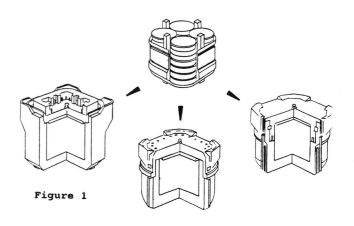

Figure 1

STANDARDISED 500 l DRUMS

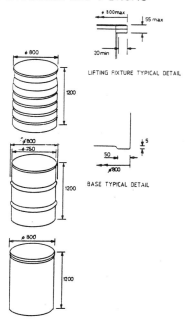

Figure 2

C431/067

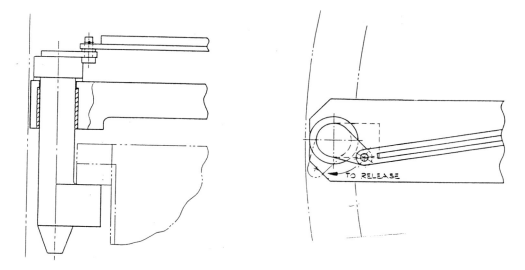

OPTION 1 Offset Tongue

Figure 3

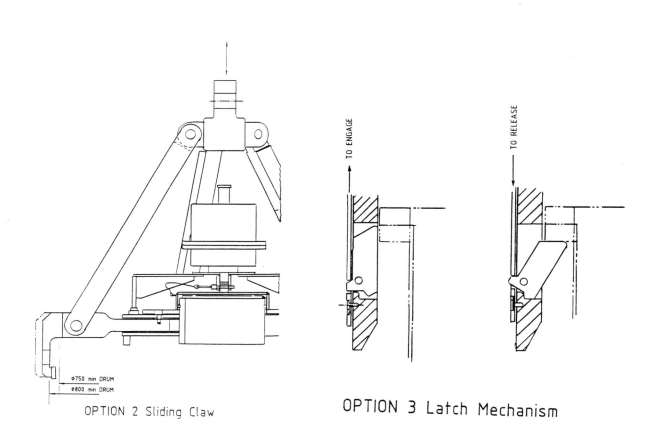

OPTION 2 Sliding Claw

OPTION 3 Latch Mechanism

Figure 4

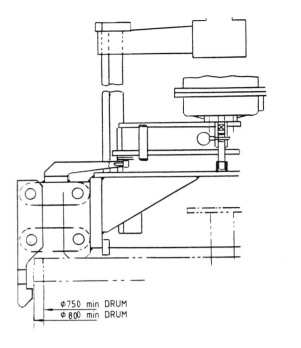

Φ750 min DRUM
Φ800 min DRUM

OPTION 4 Parallel Claw

Figure 5

OPTION 5 Parallel Claw

Φ 750 min DRUM

Φ800 min DRUM

FIG.6

500 LITRE DRUM LIFTING GRAB

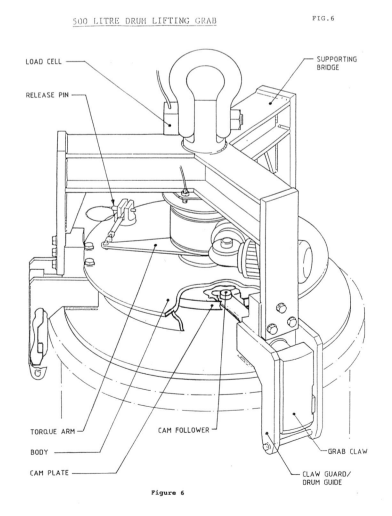

LOAD CELL

RELEASE PIN

SUPPORTING BRIDGE

TORQUE ARM

BODY

CAM PLATE

CAM FOLLOWER

GRAB CLAW

CLAW GUARD/
DRUM GUIDE

Figure 6

Development and testing of a handling system for vitrified high-level waste canisters at the Asse Salt Mine, Germany

K MÜLLER, Dipl-Ing and T ROTHFUCHS, Dipl-Geophys
GSF-Institut für Tieflagerung, Braunschweig, Germany

SYNOPSIS A pilot-test on the final disposal of high-level radioactive waste (HAW) is forseen at the Asse salt mine in Germany. Thirty vitrified highly radioactive canisters fabricated by the Battelle Pacific Northwest Laboratories (PNL), Washington – USA, will be emplaced in six boreholes at the 800-m-level below ground. A complete handling system for transportation and emplacement of the radioactive canisters has been designed and fabricated. The system was preliminarily taken into operation in 1990 in a so-called 'cold training' without any radioactive material. The handling system consists of an above ground transfer station used to unload the radioactive canisters from multiple transport casks into single transport casks which fit into the hoisting system at the Asse mine. The multiple transport casks will be also used as storage casks after retrieval of the canisters at termination of the tests. At the testing level 800 meters below the surface a specially designed transport vehicle transports the single transport casks to the test field. Here, the single transport casks are put on a so-called borehole slider which is used as a shielding valve during lowering of the canisters from the single transport cask into the emplacement borehole. The emplacement of the radioactive canisters is being performed with a remotely handled disposal machine located at the emplacement borehole. In order to guarantee the permanent retrievability of the radioactive canisters throughout the complete testing period all emplacement boreholes are lined with special steel liners. After the emplacement of five radioactive canisters the boreholes are closed for shielding purpose with a special shield plug located at top of the borehole liner.

1 INTRODUCTION

The concept applied in Germany for the final disposal of high-level radioactive waste (HAW) packages from reprocessing plants involves the emplacement in boreholes with a depth of several hundred metres in salt formations.

In a final pilot test, it will be demonstrated that salt is suitable for use as repository medium. For this purpose, 30 specially manufactured highly radioactive test sources are to be retrievably emplaced at the 800 m level of the Asse Salt Mine for a period of five years.

The test sources consist of a steel canister containing a borosilicate glass of 60 l volume. The glass contains specified amounts of the radionuclides Cs137 and Sr90. The canisters were fabricated by PNL under the provisions of a German-American agreement, in conformance with a quality assurance (QA) programme accepted by the German approval authorities. The canister lid was welded, the cover weld was tested for leakage, and the canister thus prepared was decontaminated (1). The specification of the canisters are presented in table 1.

The German repository concept makes no provision for the retrievability of waste packages. Since approval is subject to special legal stipulations at the Asse Salt Mine, however, the retrievability of the test sources must be ensured by the installation of a borehole liner for this test. During the test, therefore, the borehole liner represents real conditions as encountered with genuine HAW canisters in the final repository with respect to diameter (about 430 mm), surface temperature (maximum: 230°C), and surface dose rate. A detailed description of the overall project is presented in (2).

2 COMPONENTS OF THE HANDLING SYSTEM

Although equipment for handling radioactive components is present in various nuclear plants, it is not applicable to the conditions which prevail during routine operations underground. Hence, on the basis of experience gained from earlier experiments, almost all components required for the test had to be newly developed. The safety requirements imposed by transport regulations on the transport casks employed, the respective safety requirements of the governmental authorities competent for approval of the test performance in the Asse Salt Mine, on the one hand, and later interim storage of the test sources at an interim repository, on the other hand, had to be taken into account.

The handling system developed for the HAW emplacement test is illustrated in figure 1. The essential, newly developed components are described in the following.

2.1 Multiple transport cask (MTC)

The radioactive canisters are delivered in the MTC. They are transferred to single transport casks (STC) at a transfer facility above ground (see figure 2).

The MTC from the Castor family accommodate five canisters each. The casks are equipped with cooling fins for ensuring adequate heat dissipation. Since the MTC must also be employed for interim storage of the canisters after termination of the test, they have been designed in conformance with the requirements on storage casks. The additional requirements concern the lid area: The casks must be equipped with a double lid system and long-term metal sealings for storage.

In the six MTC, all thirty radioactive canisters can be accommodated after the end of the test.

2.2 Single transport casks (STC)

Two STC have been provided for the test. These casks have been constructed in compliance with the IAEA regulations and have been approved for shipping as a type B(U) package. They are equipped with a shield against neutron radiation, even though the test sources do not emit neutron radiation. The STC are employed for internal company transport. The canisters are drawn into the cask or withdrawn from same through an opening in the bottom. A canister grapple integrated into the lid area is employed for this purpose (see section 2.3), while the bottom opening is closed by a bottom slide.

2.3 Grapple system

The grapple system is provided for handling the radioactive canisters. This system is subdivided into two sections: So-called coupling grapples are attached to the cables of the crane facility in the transfer hall above ground as well as to those of the disposal machine underground. These grapples are equipped with windings for electromagnetic actuators and can be coupled with the so-called canister grapples in the head of the STC.

2.4 Transfer facility above ground

The transfer facility has been installed in a special newly constructed hall. It comprises the transfer station and the crane facility (see figure 2). The crane facility consists of a bridge crane with a trolley and is equipped with three hoisting devices for raising and lowering the STC, for placing a special shielding bell on the STC, and for manipulating the radioactive canisters. The transfer station is employed for docking an MTC, for removing the primary lid from the MTC with the shielding maintained, as well as for opening and closing the MTC in position. The crane fa-

cility is controlled and monitored by means of a stored programme control system (SPS); the transfer station is provided with contactor control. Interlocks, antivalence monitoring, and redundancies ensure safe operation for the personnel and preclude all possibilities of faulty operation relevant to protection against radiation.

2.5 Transport vehicle

The transport vehicle (see figure 3) is employed for ground transport of the STC from the shaft to the test location and back. It is equipped with a hoisting device for raising and lowering the STC. The hydraulic pump for the travel and hoisting drive systems is driven by a Diesel engine. The vehicle has been designed with centre pin steering; the loading platform on the front carriage is equipped with devices for securing the STC and protecting it in case of an accident. A fire-fighting facility for use in case of a tyre fire, for instance, is permanently installed on the vehicle.

2.6 Borehole slider and borehole plug

The borehole slider provides the necessary shielding against radiation during emplacement or retrieval of the canisters and also accepts the STC, while the borehole plug provides radiation shielding after removal of the borehole slider. The borehole plug is lowered in the same way as a canister from the STC before dismounting of the borehole slider and closes the borehole.

2.7 Disposal machine

The disposal machine (see figure 3) is employed for emplacement and retrieval of the canisters from the boreholes. It can be repositioned, since six boreholes must be served.

The disposal machine consists mainly of the winch unit, a control unit, and the undercarriage. The winch is driven by an electric motor; in the event of a mains power failure, power is supplied by integrated batteries. The machine is movable in all directions and is positioned at each borehole in such a way that the winch is in a direct vertical position above the borehole slider. It remains in this position until charging of a borehole has been completed. During placement of the STC on the borehole slider, the winch unit of the disposal machine is swung aside.

All manipulations with canisters and the borehole plug are performed with this machine. Furthermore, it controls the borehole slider and shielding bell equipped with position indicators and drive units.

Since the machine has been developed as a prototype for a final repository, the hoisting winch has been designed for a borehole depth of 300 m, although the boreholes involved in the test are only 15 m deep. Regardless of the borehole depth, the following special conditions had to be taken into account:

116

(a) voltage drop during the operation of electromagnets, as a result of the cable length;

(b) consideration of the cable weight as a function of the grapple depth with or without payload.

The handling operations are controlled by means of a stored programme control unit in such a manner that the respective final positions of the grapple or load are approached with caution, independently of any intervention by the operating personnel. Moreover, the respective emplacement positions of the canisters are stored in the memory during emplacement; thus, the canisters can be manipulated with the appropriate care during retrieval.

2.8 Borehole liner

Lining of the disposal boreholes is necessary for ensuring retrievability of the canisters at any time. Retrievability is a special requirement imposed by the governmental approval authority for tests involving radioactive materials in the Asse salt mine.

Because of the scientific objectives of the test, two different types of borehole liners have been installed:

(a) In the case of type A, the annular gap between the salt and the liner is filled with poured corundum spheres, which prevent a tight enclosure of the liner in the salt.

(b) In the case of type B, the creeping salt converges around the liner within a short time and encloses it firmly. Release of volatile components from the salt is thus largely avoided, in contrast to type A.

The design, manufacture, and installation, including the required quality assurance, presented an engineering challenge (3). Only a few highlights are mentioned here:

(a) The lowermost section of the borehole liner had to be thin-walled, and therefore manufactured of ultrahigh-strength steel. A Maraging steel was employed for the purpose. Machining of such large workpieces made of this steel grade was an entirely new problem. Moreover, testing of the workpiece had to be planned and executed for ensuring quality in all details.

(b) The problem of corrosion protection has been solved by complete, gas-tight jacketing of the liner surface with TiPd plate 3 mm thick on the side exposed to the salt.

(c) For accepting and guiding installed objects inside the borehole liner (a canister guiding system and a gap monitoring system), rails had to be welded in place very accurately. A welding facility had to be developed for this purpose. Qualification of the welds was necessary.

(d) The deformation of the 'weakest' section of the borehole liner must be monitored. An oval deformation is assumed in the middle of the canister stack. The systems provided for the purpose had to be designed and tested for reliability before the beginning of manufacture. The development of the deformation monitoring system was assumed by the Dutch partner in the project, ECN in Petten (4).

3 DESCRIPTION OF CANISTER HANDLING

In the following, the emplacement is described with the use of the flow chart presented in figure 4. The retrieval operation is illustrated diagrammatically in figure 5.

For the sake of clarity, a specified sequence is assumed in figure 4; deviations from this sequence are possible in reality.

The first loop is repeated until all thirty canisters, that is, all six MTC, have arrived at the mine, since shock absorbers are available for only two casks.

After unloading of the MTC from the car, the MTC to be emptied must be prepared for further actions. These comprise, in particular: removal of the primary lid from the MTC, removal of the cover screws for the secondary lid, insertion of guide bolts, and positioning of the lid adapter. The MTC is subsequently pushed under the transfer station (block C).

The MTC is emptied with the use of both STC, which travel in a shuttle operation between the emplacement borehole underground and the transfer facility above ground (blocks D and E) until the borehole has been filled with five radioactive canisters.

The transfer operation above ground consists of many individual steps, which in part proceed fully automatically. Other portions of the individual steps are started at the press of a button and are automatically terminated after execution in each case. A few manipulations must be controlled manually. In any case, safety-relevant steps are mutually interlocked, even if the automation is switched off.

The transfer operation begins with the placement of the STC on the transfer station. The coordinates of the point from which the STC has been taken is thereby stored in the memory; the completed transfer is followed by a return to this point. The positioning point of the STC on the transfer station is also stored and is approached very exactly. The speed at which the load is raised and lowered can be continuously selected by the crane operator within a wide range. A direct view of the installations and a monitor facilitate the operation of the crane. After release of the STC from the crane hook, the STC is moved horizontally through a distance of 958 mm to a position over the MTC. Since the running points on the cables of the 10 t hoisting device and of the winch for hoisting the canisters are likewise separated by a distance of 958 mm, no travel of the bridge crane is necessary. The advantage of this feature is the avoidance of possible swinging motion of the loads and hook. The following operational steps proceed in succession after attainment of the transfer position:

(a) Coupling of the coupling grapple on the canister in the STC
(b) Setting the shielding bell in place
(c) Selection of the canister to be transferred by rotation of the MTC
(d) Pulling of the locking bolts for the bottom slide in the base of the STC
(e) Opening of the bottom slide in the STC
(f) Localization of the grapple on the canister situated in the MTC, grasping of the canister, and lifting into the STC
(g) Closing of the bottom slide and mechanical locking of the bottom slide
(h) Setting of the canister in place on the bottom slide and disengagement of the canister
(i) Raising of the shielding bell and release of the canister grapple from the coupling grapple
(j) Horizontal return travel of the STC over the distance of 958 mm mentioned to the position from which the STC can be withdrawn again; no motion of the crane required for this purpose

A heavy-duty flat-bed trailer with a tractor is available for transporting the STC from the transfer hall to the hoisting shaft. In the shaft hall above ground, the route of the empty STC arriving from the underground installations intersects that of the filled STC. A fork lift has been specially converted for the purpose of transfer from the heavy-duty trailer to the hoisting cage and vice versa.

The underground activities begin with the following preparatory operations for emplacement:

(a) Preparation of the borehole for receiving the radioactive canisters; termination by placement of one of the two available borehole sliders;
(b) Positioning of the disposal machine on one side of the borehole concerned; pay-off point of the cable for canister handling exactly coaxial with the borehole.

The loaded STC arriving from the facility above ground is removed from the hoisting cage by the hoisting device of the transport vehicle and placed on the borehole slider at the borehole. During the ground transport operation, the STC is secured against accidents on the platform of the vehicle.

The emplacement of the radioactive canister in the borehole closely resembles the canister transfer above ground:

(a) Coupling of the coupling grapple on the canister in the STC
(b) Setting the shielding bell in place
(c) Grasping of the canister and removal from the bottom slide
(d) Pulling of the locking bolts for the bottom slide in the base of the STC
(e) Opening of the bottom slide
(f) Lowering of the canister into the borehole, initially at low speed
(g) Decrease in speed before attainment of the emplacement position (because of progressing load at different depths)

(h) Setting of the canister in place and disengagement of the canister (storage and processing of the depth for the next emplacement operation and for later retrieval)
(i) After retraction of the grapple into the STC, closing of the bottom slide and mechanical locking of the bottom slide
(j) Raising of the shielding bell and release of the canister grapple from the coupling grapple

The STC can now be lifted with the use of the transport vehicle.

The so-called dummy canister is mentioned in block F. This canister has the same external dimensions as a radioactive canister. The dummy canister is employed for an irradiation programme with salt samples and for decreasing the radiation in the direction of the borehole mouth. The dummy canister is inserted into the system by raising the lid of the STC. The borehole plug is inserted in the same manner. In spite of its large diameter, it can be handled with the disposal machine by means of the STC and borehole slider.

The retrieval sequence is illustrated in figure 5 and is the reverse of the emplacement procedure.

3   OPERATING EXPERIENCE AND PROSPECT

Only test runs have hitherto been conducted (as of April 1991) with the facilities and components with the use of inactive canisters. These tests have demonstrated that safe operation is feasible. Weaknesses have been eliminated, and the organization of the emplacement as well as of the retrieval has been improved. It has been shown that the so-called quick retrieval of all thirty canisters to the surface can be accomplished in less than fifty hours. Especially the vehicle manipulations and switching operations have proved to be relatively time-consuming. These operations will be the object of training in the future.

The emplacement of the thirty high-level radioactive canisters is at present not possible because of the laws relating to approval. In view of the overall schedule with respect to the construction and commissioning of a German final repository for radioactive wastes, however, the emplacement should be conducted before the mid-1990's, with a test duration of five years.

Acknowledgement

This project was performed in close cooperation with the Netherlands Energy Research Foundation and mainly funded by the German Minister for Research and Technology and the Commission of the European Communities in the framework of the Community Research Programme.

REFERENCES

(1) HOLTON, L. K., BURKHOLDER, H. C., McEL-
ROY, J. L., KAHL, L., KROEBEL, R., ROTH-
FUCHS, T., STIPPLER, R. Fabrication of
vitrified isotopic heat and radiation
sources for testing in the Asse mine. In:
Proceedings Waste Management 89, Tucson,
Arizona/USA, Feb. 26 – March 02, 1989
(Eds. Post/Wacks), Arizona Board of
Regents/USA, Vol. 1, pp. 575 – 586, 1989.

(2) ROTHFUCHS, T., STIPPLER, R. High-Level
Radioactive Waste Test Disposal Project
in the Asse Salt Mine – Fed. Rep. of
Germany. In: Proceedings Waste Management
87, Tucson, Arizona/USA, March 1-5, 1987,
(Eds. Post/Wacks), Arizona Board of Re-
gents/USA, Vol. 1, pp. 89 – 95, 1987.

(3) SCHÄFER, P., MÜLLER, K. Hochfeste Bohr-
lochverrohrung zur versuchsweisen Einla-
gerung von HAW-Glasblockkokillen im Salz-
bergwerk Asse. In: Jahrestagung Kerntechnik '90, (Hrsg.: Dt. Atomforum e. V.,
Bonn), S. 325 – 328, 1990.

(4) DE BOER, I., VRIESEMA, B., ROTHFUCHS, T.
Engineering Safety Aspects of the
High-Level Radioactive Waste Test Dis-
posal Project in the Asse Salt Mine. Pro-
ceedings of the International Symposium
of the Back-End of the Nuclear Cycle –
Strategies and Options, Vienna, Austria,
May 11 – 15, 1987, IAEA-SM-294/66, 1987.

Table 1 Specification of the highly radioactive canisters

Height: 1154 mm; outside diameter: 298,5 mm; content of borosilicate glass: 60 l

Type	Average heat output		Nuclide inventory		Maximum salt temperature	Average gamma dose rate (in air)
	W/canister	W/l			°C	kGy/h
I	1335	22.0	Sr 90	(Cs 137)	160	0.93
II	1490	25.0	Cs 137	Sr 90	180	2.28
III	1860	31.0	Cs 137	Sr 90	230	2.51

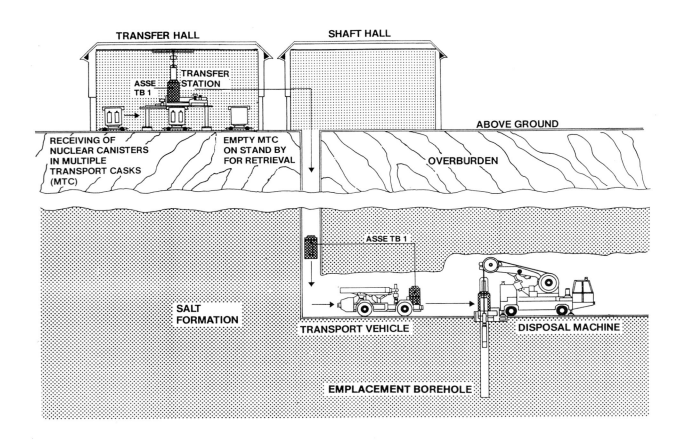

Fig 1 Transport and emplacement system

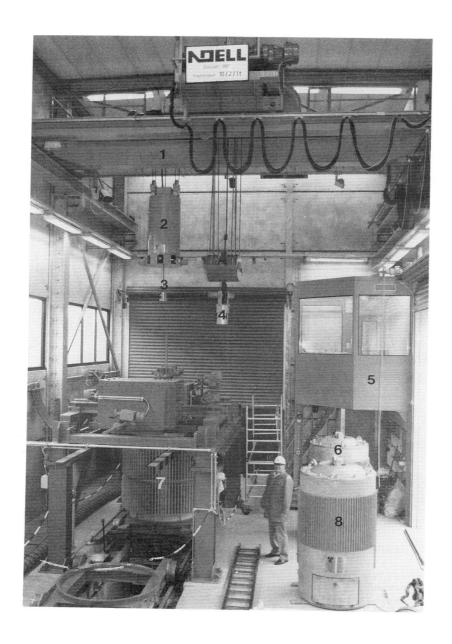

Fig 2 Above ground transfer hall with transfer station and crane at start of unloading procedure

1. crane, 2. shielding bell connected to the 2 t winch, 3. suspension clamp (coupling grapple disassembled), 4. load hooking device connected to the 10 t winch, 5. control stand, 6. canister grapple, 7. multiple transport cask (MTC), 8. single transport cask (STC)

Fig 3 Emplacement system at an emplacement borehole

1. transport vehicle, 2. disposal machine, 3. shielding bell connected to the 2 t winch of the disposal machine, 4. load hooking
device, 5. single transport cask Asse TB1 (STC) deposited onto the
borehole slider

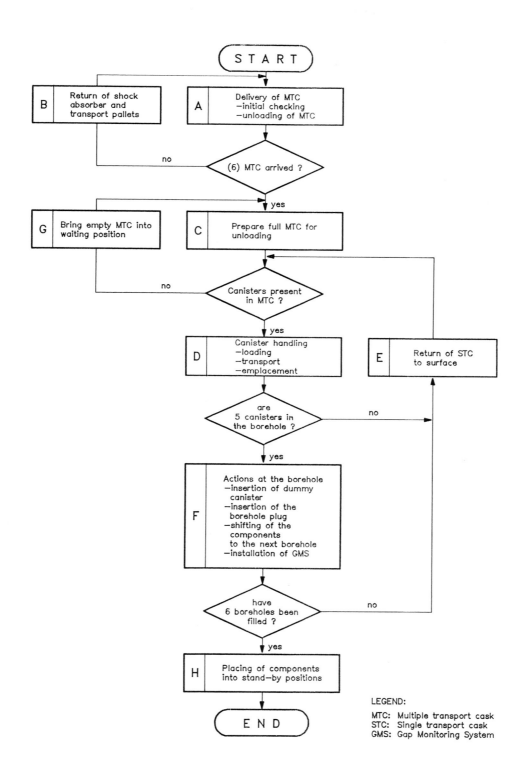

Fig 4 Flow chart of HAW emplacement

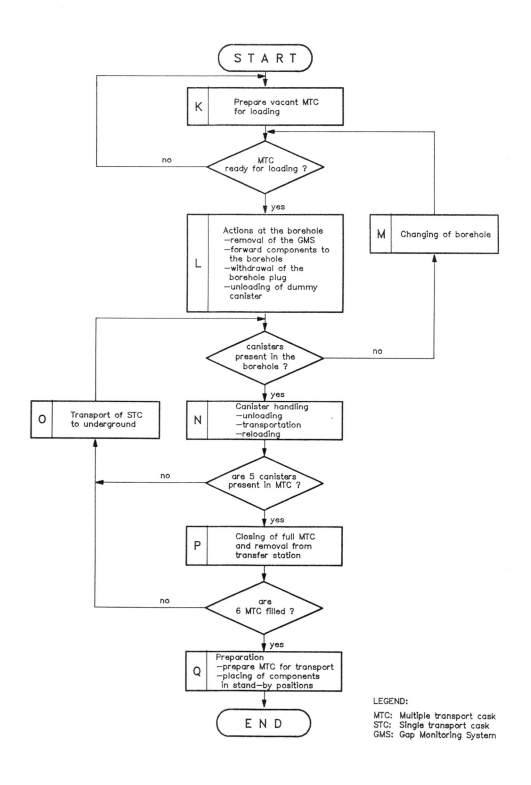

Fig 5 Flow chart of HAW unloading operations

Mechanical handling systems in the Sellafield vitrification plant

S M HANSFORD, BSc, CEng, MIMechE and D R COULTHART, BSc
British Nuclear Fuels plc, Risley, Warrington, Cheshire
J W D MILLER, BSc
British Nuclear Fuels plc, Sellafield, Seascale, Cumbria

SYNOPSIS   British Nuclear Fuels plc (BNFL)has over 40 years experience in the design, construction and operation of nuclear reprocessing plants and of waste management. Many of these plants have required extensive mechanical handling systems, the handling and control systems designed and developed for the Sellafield Vitrification Plant and Product Store are described.   These systems are now fully operational and illustrate many of the features and techniques developed by BNFL for nuclear package handling. Utilisation of these systems and similar systems in other Sellafield Plants has demonstrated notable advantages in ease/flexibilty of operations, product quality and costs.

1   INTRODUCTION

Sellafield is BNFL's centre for nuclear reprocessing and waste management operations. Major capital programmes are underway to extend and improve these facilities. Design and development work for many of these programmes began in the early 1980's and continued through into the 1990's as part of the progressive expansion of the Company's facilities. Some £4000 million of new plant will be commissioned and brought into operation at Sellafield during the early 1990's.

The major plants included in this expansion of Sellafield's facilities are :-

-    Thermal Oxide Reprocessing Plant (THORP).

-    Encapsulation Plant No. 1 and Product Store (EP1 and EPS1).

-    Encapsulation Plant No. 2.

-    Vitrification Plant and Product Store.

These new plants have required reliable and safe radioactive waste handling and control systems to meet the high standards necessary for nuclear plant operation.

The Sellafield Vitrification Plant and its associated Product Store is one of these plants. Design, construction and commissioning were carried out by BNFL. It was brought into operation in mid 1990; capital costs were £240 million.

The Vitrification Plant converts highly radioactive liquid waste (HLW) into a stable solid form. A continuous process is used in which HLW is dried and mixed with glass forming additives, melted and poured into stainless steel containers. After cooling, lid welding, decontamination and monitoring, the containers are stored in an engineered Product Store.

There are two process lines in the plant. Each line comprises a Vitrification Cell, Breakdown Cell and a Pouring Cell. A common Decontamination Cell plus a Control Cell serve both process lines.

The plant and its associated product store has the capacity to treat and store future arisings from reprocessing operations; and the backlog of stored HLW.

2   MECHANICAL HANDLING AND CONTROL SYSTEM

The mechanical handling and control system caters for the processing of product containers through the complete vitrification cycle, ie. from the introduction of clean product containers through to the placement of filled, welded and decontaminated containers in the Product Store.

The mechanical handling system is made up of the following equipment:-

i.   5 off in-cell polar cranes - one in each Vitrification/ Breakdown Cell; one in each Pouring Cell and one in the Control Cell.

ii.   3 off product flasks - for the transfer of containers from the Vitrification Plant into the Product Store.

iii.   2 off charge flasks - for preparing a storage tube to receive product containers.

iv.   1 off transfer bogie - to transport product flasks from the

Vitrification Plant to the Product Store and vice versa.

v.   1 off 50 tonne plant crane - for handling product flasks onto/from the transfer bogie in the Vitrification Plant.

vi. 1 off 50 tonne store crane - for handling charge flasks, product flasks and associated equipment.

Control of the mechanical handling equipment is from local control desks in the operating areas. At these desks all equipment control and signalling devices are present. Data from these desks is relayed to the central control room (CCR) and the supervisory computer system.

Interfacing with these control desks are 21 programmable logic controllers (PLCs) dedicated to ensuring the safe operation of the mechanical equipment. These PLCs oversee operations within the cells interfacing with one another in some cases, to allow multiple operations. They are based upon available proprietary processors and are designed to operate without air conditioning or electrical supply filtering (see Figure 4).

All of the above equipment has been designed and manufactured to exacting BNFL engineering standards to ensure:-

-   fitness for purpose

-   reliability

-   cost effectiveness

-   product quality

-   safety

3   VITRIFICATION AND BREAKDOWN CELLS

In the Vitrification Cell, HLW is dried in a calciner, mixed with glass forming additives and vitrified in the melter. This molten product is then poured into a product container at the segregation dome in the Pouring Cell.

Equipment used in the Vitrification Cell requires replacement from time to time. This equipment is transferred into an adjacent Breakdown Cell where it is size reduced, characterised and packaged. Major items of equipment which will be removed periodically from the Vitrification Cell includes:-

-   melter crucibles

-   melter inductors

-   calciner tubes

-   dust scrubbers

The Breakdown Cell is separated from the Vitrification Cell by a short passage. An in-cell crane is provided and is mounted on rails running the full length of the Vitrification and Breakdown Cells. This crane is similar to those used in the Pouring and Control Cells. Crane maintenance and decontamination facilities are located at the South end of the cell complex. Through wall telescopic master slave manipulators (MSMs) and shielded viewing windows are provided at strategic positions in the cells.

Equipment available within the Breakdown Cell includes:-

-   mobile shear

-   reciprocating saw

-   bench

-   decontamination facility

-   monitoring facility

-   plasma arc welder

The mobile shear unit is deployed by the in-cell crane. It is used to shear integral pipework and supports from equipment. It can also be deployed in the Vitrification Cell as required. Pipework up to 80mm diameter can be sheared.

The reciprocating saw provides the main means of size reducing equipment. Items to be cut are clamped by chains allowing equipment of various sizes and shapes to be secured. After cutting, sections are removed using the in-cell crane fitted with a grapple and deposited into a waste basket.

Waste baskets are mounted in the cell bench. Baskets can be decontaminated in the decontamination tank. After decontamination the baskets are monitored for radioactivity to determine the appropriate storage route.

High active (HA) waste baskets are placed in a container and a lid is welded in place. These containers are despatched to the Product Store (see Figure 3).

Medium active (MA) waste baskets are posted out of the cell into a transit container and taken to the site Miscellaneous Beta Gamma Waste Store.

All maintenance operations in the Vitrification and Breakdown Cells are carried out remotely by trained operators. Emphasis is placed on the training of maintenance operators and the use of tried and tested maintenance procedures to ensure that all operations are carried out in a safe and efficient manner.

Control of maintenance operations is from local control panels at the cell face. These are located below shielded viewing windows with additional cell viewing provided via monitors and in-cell cameras.

In-cell crane control is via a PLC based system. All crane operations are semi-automatic under operator control.

4 POURING CELLS

The Pouring Cells are located directly beneath their respective Vitrification Cells (see Figure 1). Cell dimensions are approximately 23.0m long x 4.0m wide x 5.5m high. Radiation levels within the cell can be up to 10^2 Gy/hour.

All operations are carried out remotely and are controlled by operators working at local control desks on the cell face

e operations carried out are:-

i. tation of an empty container at the filling station
e molten glass product from the melter.

ii. the filled container from the filling head
id on the container.

iii. tainer to a cooling station.

iv. r from the cooling station to the

v.

vi.

into the Decontamina-

transfers between stations through to posting out into the Decontamination Cell is carried out by the in-cell crane.

5 IN-CELL CRANE

The in-cell cranes are arranged in polar form as shown in Figure 2. In the design of these cranes attention has been focussed on achieving a high degree of reliability, methods of recovering, safely and quickly from a failure and ease of maintenance.

The cranes are constructed in stainless steel and have a modular arrangement. Rated capacity is 2 tonne. Each crane motion has dual drives; main and back-up. The hoist unit is a two fall link chain. Product containers are grappled via an electrically actuated finger grab.

Crane main drives are variable speed with programmable positional and velocity control. The back-up drives are via single slow speed a.c. motors.

Power feeds and control signals are via dual downshop cables from reeling drums located in the crane maintenance area.

Control of the crane is via its own PLC with automatic and semi-automatic modes. The crane control PLC interfaces with the cell work station PLCs; zoning of the control systems allows more than one operation to take place in the cell at any one time, provided the safety interlock conditions are met.

Crane maintenance and decontamination facilities are located at the South end of the cells. Here crane modules can be remotely removed and replaced through the provision of MSM's and a jib crane. Modules may be remotely repaired in the crane maintenance area. Alternatively after decontamination modules can be repaired by hands-on techniques in a dedicated workshop.

6 DECONTAMINATION CELL

A decontamination cell common to both vitrification lines contains a decontamination vessel which traverses on rails from the post down facility for each pouring cell and each Breakdown Cell to a post up facility in the Control Cell. Inflatable seals are used at each posting position to seal the decontamination vessel to the cell port.

The in-cell crane (from the cell above) lowers the container through a trap door onto a turntable in the decontamination vessel. Here high pressure water or dilute nitric acid solution is used to remove any loose contamination from the xternal container surfaces.

127

Control of the decontamination cycle is via a PLC which controls the decontamination sequences and interfaces with the PLCs of the in-cell cranes in the Pouring, Breakdown and Control Cells. This interface ensures safety interlocking of the cell trap doors so that they cannot be operated unless the decontamination vessel is in position and a decontamination cycle is not in progress.

A secondary role for this facility is to decontaminate containers of highly active waste generated from operations in the Breakdown Cell enroute to the Product Store.

7   CONTROL CELL

Welded, decontaminated product containers and HA waste containers are transferred from the Decontamination Cell into the Control Cell. Here, the external container surfaces are swabbed to confirm that effective decontamination has been carried out prior to export to the Product Store.

The cell is equipped with an almost identical in-cell crane to that used in the Pouring Cells. This crane is used to lift the container into the cell from the Decontamination Cell and for all other container transfers within the cell.

Containers are placed on a turntable and rotated. A programmable robot is used to deploy a swab over the entire external container surface. Swabs are transferred via a pneumatic transport system to an external monitoring station.

After confirmation that the container meets stringent acceptable limits for external contamination the container can be posted from the cell into a product flask for transfer to the Product Store.

Crane control and maintenance facilities are similar to those provided in the Pouring Cell.

8   CONTAINER EXPORT

Containers are exported from the Vitrification Plant in a shielded product flask. The sequence of operations involved in the export of containers is shown in Figure 3, and basically comprises:-

-    positioning a product flask on the Control Cell export gamma gate, using the plant crane.

-    hoisting a container from the Control Cell into the flask using the flask hoist.

-    transferring the product flask from the Control Cell gamma gate to the transfer bogie using the plant crane.

-    transfering the product flask on-board the transfer bogie into the Product Store via the air lock.

To ensure safe operation, all actions are interlocked via the PLC. Electro/mechanical locking features are provided on the flask doors.

9   PRODUCT FLASK

A shielded product flask is used to transfer product containers and HA waste containers from the Control Cell to the Product Store.

These flasks, of which there are 3 in-service, provide full radiation shielding of containers during the transfer operations. Flask weight is 39 tonne, overall dimensions are 3.6m high, with a base of 1.62m x 1.58m. Materials of construction are ductile cast iron and carbon steel incorporating borosilicate neutron shielding materials. Integral to the flask is a hoist unit with a rated capacity of 670 kg. An electrically actuated finger grapple is used to lift containers, similar to that used in the pouring cell (see Figure 5).

10   PLANT CRANE

An electrical overhead travelling crane is provided product flasks between the control cells and th bogie. The crane has a rated capacity of 50 comprises a hoist unit and carriage, i.e. all lift a points are in a straight line.

In common with all the mechanical han used in the plant, careful attention has been and manufacture of the crane to ensure integrity and reliability is achieved. Fe crane include :-

-    load sensing

-    seismic qualification

-    rope drum overspeed senso

-    full life time quality record bility

-    special purpose design

-    detail design of all co permissible working

Control of the cra controller.

11   TRANSFER BOGIE

The transfer bogie is of conventional design with a load platform having location features for the product flask. It is a twin axle bogie with four wheels running on rails between the Vitrification Plant and the Product Store. One axle is driven by a variable speed a.c. motor and speed reducing gearbox; power supply and control feeds are via a trailing cable paid out/taken up from an on-board cable reeling system.

Control of bogie movements is by an operator via a pendant controller/deadmans handle. Interlocking for safety purposes is through the PLC.

12   PRODUCT STORE

The Product Store is adjacent to the Vitrification Plant and is some 40.0m high x 30.0m wide x 53.0m long. It comprises four compartments, each having 200 vertical double walled thimble tubes. Product containers and HA waste containers are stacked vertically up to 10 high in each thimble tube (see Figure 6).

Heat generated by the container product is removed from the thimble tubes by a cooling air flow through the annular space between the two tubes. This air flow is generated by natural convection. Since under normal operation there is no direct contact between the cooling air and the containers, there is no danger of circulating cooling air containing radioactive particulate. Hence there is no requirement for air filtration prior to discharge to atmosphere.

A 50 tonne overhead store crane is provided to cover the whole charge floor area. This crane is used for all handling operations on the charge floor.

Before containers can be transferred from the product flask into a storage thimble, the selected thimble requires preparation, eg. removal of the floor plug, thimble cap and the insertion of a charge chute.

This preparation basically involves the following operations:-

- locate and fasten an adaptor onto the floor plug immediately above the selected thimble.

- locate a gamma gate onto the adaptor.

- position No. 1 charge flask onto the gamma gate and remove the floor plug.

- position No. 2 charge flask onto the gamma gate, insert a charge chute and remove the thimble tube cap.

The product flask can now be positioned on the gamma gate using the store crane and a container emplaced in the thimble tube. The sequence of operations involved are summarised in Figure 3 and basically include:-

- opening the flask and gamma gate doors.

- lowering the container into position in the thimble tube.

- raising the container grab into the flask.

- closing the flask and gamma gate doors.

When the selected thimble tube is filled with containers, to a maximum of 10 high, the sequence of preparation operations outlined above is reversed, ie. thimble cap replaced, charge chute removed and floor plug replaced.

A series of electro/mechanical detectors and switches are built into the gamma gate and flasks. These are interfaced to a PLC to ensure a safe series of operations are carried out. The position of each container is logged into the Product Store supervisory computer.

Environmental monitoring and recording of the store is carried out including air flow rates and temperatures.

A facility to export containers from the store will be provided. This is currently in design and construction. It will be located adjacent to the store to allow containers to be loaded into a suitable transport flask.

The facility will comprise of a Control Cell, similar to the Vitrification Plant Control Cell complete with in-cell crane and swabbing robot. A flask handling crane will also be provided and a road/rail terminal.

13   STORE CRANE

The store crane is an electrical overhead travelling crane with long travel, cross travel and hoist motions. It is rated at 50 tonne and has a span of 18.75m.

In common with the Vitrification Plant crane careful attention has been paid to the design and manufacture of this crane to ensure a high standard of integrity and reliability is achieved. The features built into this crane are similar to those listed for the Vitrification Plant crane.

Power feeds and control signals are supplied by a festooned cable system running parallel to the downshop rails. Control of the crane is by the operator via a pendant controller. Interlocking and zoning of the crane is via the PLC.

14 CHARGE FLASK

Shielded Charge Flasks are used to prepare the selected storage thimble to receive containers.

These flasks are made from carbon steel with concrete infill. Flask weight is 29tonne, dimensions are 4.58m high with a base of 1.62m x 1.58m. An integral hoist unit is included, rated at 1350kg.

15 QUALITY ASSURANCE

The Vitrification Plant and Product Store management and working procedures comply with BS 5882 : `Specification for A Total Quality Assurance Programme for Nuclear Installations'.

Activities are regularly audited both internally by BNFL's Quality Assurance Department and by external auditors. Product quality, agreed against a residue specification with BNFL's customers is assured by strict control of all process inputs, working procedures and inventory records.

16 CONCLUSIONS

The mechanical handling equipment and systems for the Vitrification Plant and Product Store were designed, manufactured and installed to high quality standards within a demanding programme of plant design and construction. Commissioning and operation of the plant has proved the functionality and reliability of the handling equipment and systems.

The equipment and systems described are specific to the Vitrification Plant and Product Store. However, they illustrate many of the features and techniques developed by BNFL for nuclear package handling. Similar equipment, systems and techniques are widely used in Sellafield plants.

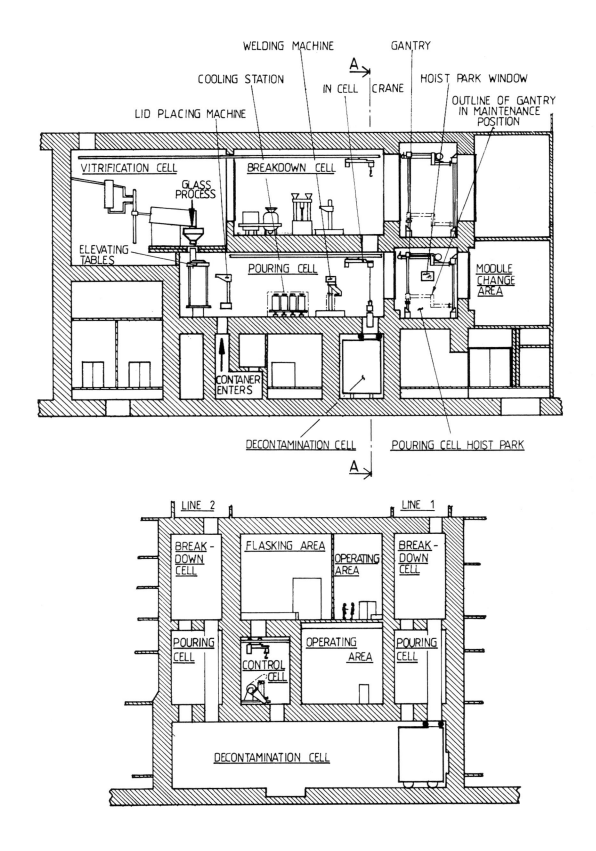

WELDING MACHINE

GANTRY

COOLING STATION

IN CELL CRANE

HOIST PARK WINDOW

LID PLACING MACHINE

OUTLINE OF GANTRY
IN MAINTENANCE
POSITION

A

VITRIFICATION CELL

BREAKDOWN CELL

GLASS
PROCESS

ELEVATING
TABLES

POURING CELL

MODULE
CHANGE
AREA

CONTAINER
ENTERS

DECONTAMINATION CELL

POURING CELL HOIST PARK

A

LINE 2

LINE 1

BREAK-
DOWN
CELL

FLASKING AREA

OPERATING
AREA

BREAK-
DOWN
CELL

POURING
CELL

CONTROL
CELL

OPERATING
AREA

POURING
CELL

DECONTAMINATION CELL

SECTION A-A

FIG 1 – VITRIFICATION PLANT CELL LAYOUT

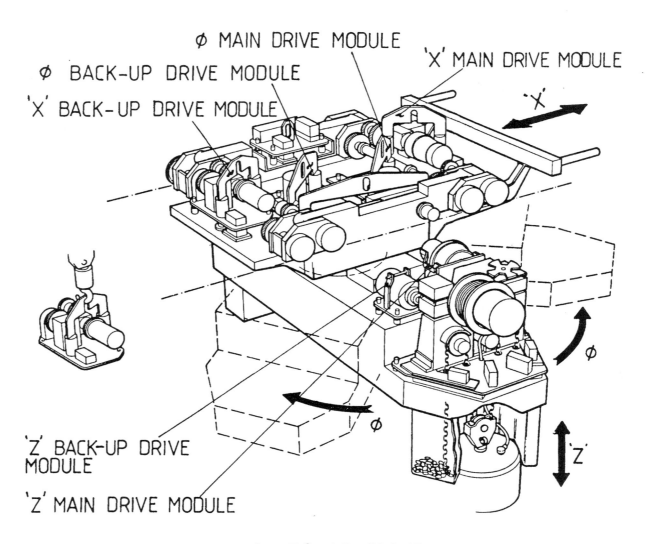

Ø MAIN DRIVE MODULE

Ø BACK-UP DRIVE MODULE

'X' MAIN DRIVE MODULE

'X' BACK-UP DRIVE MODULE

'X'

Ø

Ø

'Z'

'Z' BACK-UP DRIVE MODULE

'Z' MAIN DRIVE MODULE

FIG 2 — POLAR CRANE

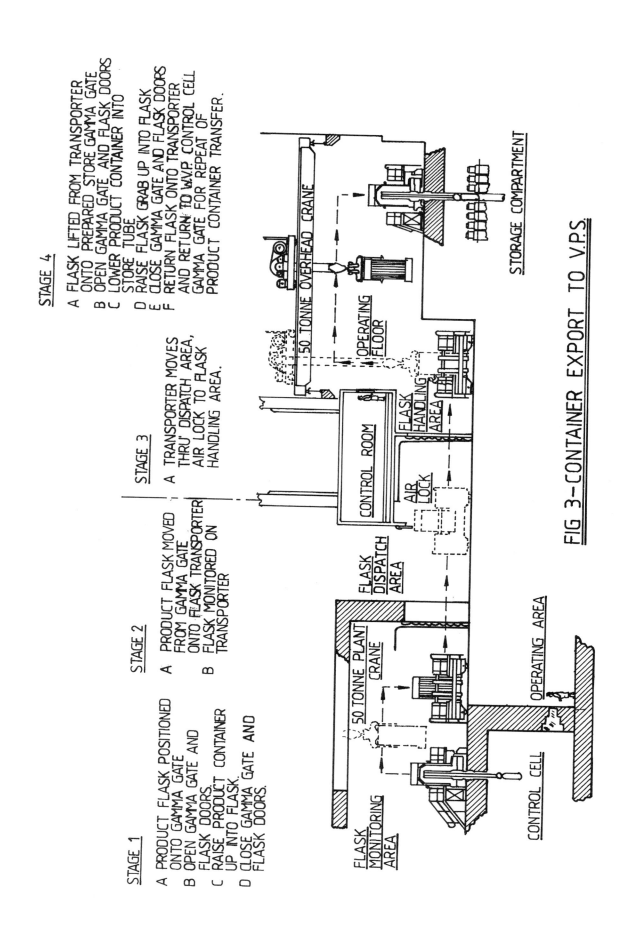

STAGE 1

A PRODUCT FLASK POSITIONED ONTO GAMMA GATE
B OPEN GAMMA GATE AND FLASK DOORS.
C RAISE PRODUCT CONTAINER UP INTO FLASK.
D CLOSE GAMMA GATE AND FLASK DOORS.

STAGE 2

A PRODUCT FLASK MOVED FROM GAMMA GATE ONTO FLASK TRANSPORTER
B FLASK MONITORED ON TRANSPORTER

STAGE 3

A TRANSPORTER MOVES THRU' DISPATCH AREA, AIR LOCK TO FLASK HANDLING AREA.

STAGE 4

A FLASK LIFTED FROM TRANSPORTER ONTO PREPARED STORE GAMMA GATE
B OPEN GAMMA GATE AND FLASK DOORS
C LOWER PRODUCT CONTAINER INTO STORE TUBE
D RAISE FLASK GRAB UP INTO FLASK
E CLOSE GAMMA GATE AND FLASK DOORS
F RETURN FLASK ONTO TRANSPORTER AND RETURN TO W.P. CONTROL CELL GAMMA GATE FOR REPEAT OF PRODUCT CONTAINER TRANSFER.

FLASK MONITORING AREA

50 TONNE PLANT CRANE

CONTROL CELL

OPERATING AREA

FLASK DISPATCH AREA

CONTROL ROOM

AIR LOCK

FLASK HANDLING AREA

50 TONNE OVERHEAD CRANE

OPERATING FLOOR

STORAGE COMPARTMENT

FIG 3—CONTAINER EXPORT TO V.P.S.

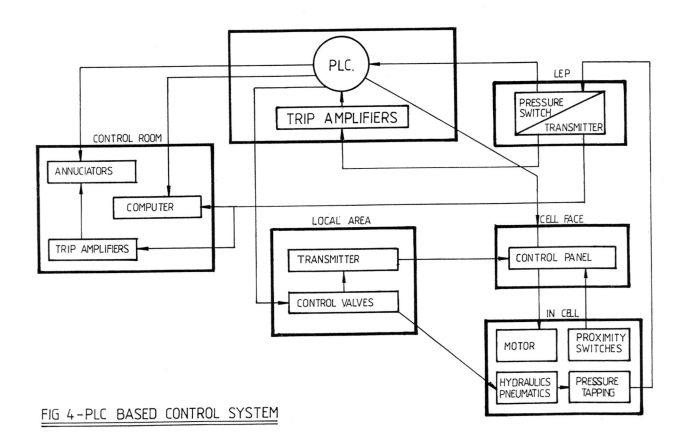

FIG 4 - PLC BASED CONTROL SYSTEM

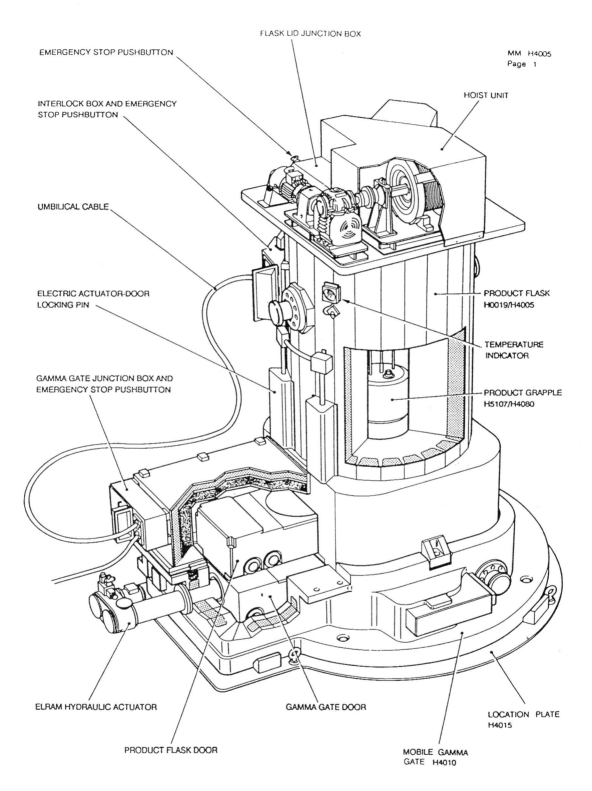

EMERGENCY STOP PUSHBUTTON

INTERLOCK BOX AND EMERGENCY
STOP PUSHBUTTON

UMBILICAL CABLE

ELECTRIC ACTUATOR-DOOR
LOCKING PIN

GAMMA GATE JUNCTION BOX AND
EMERGENCY STOP PUSHBUTTON

FLASK LID JUNCTION BOX

HOIST UNIT

PRODUCT FLASK
H0019/H4005

TEMPERATURE
INDICATOR

PRODUCT GRAPPLE
H5107/H4080

ELRAM HYDRAULIC ACTUATOR

PRODUCT FLASK DOOR

GAMMA GATE DOOR

MOBILE GAMMA
GATE H4010

LOCATION PLATE
H4015

FIG 5 - PRODUCT FLASK AND MOBILE GAMMA GATE

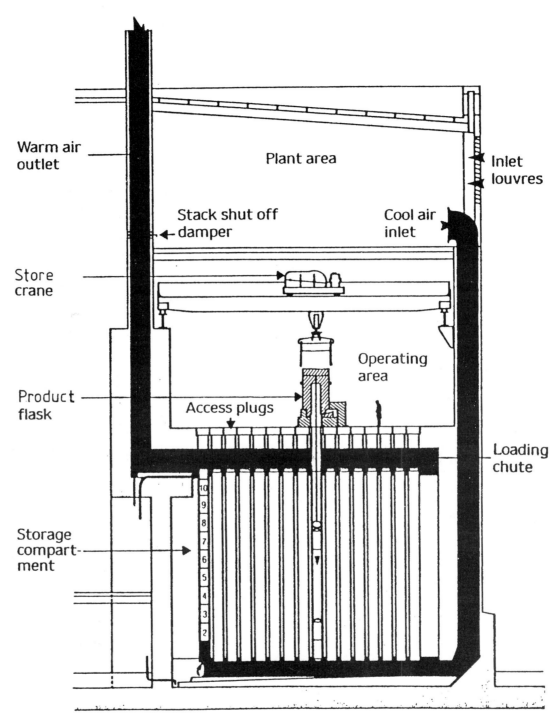

Warm air outlet

Plant area

Inlet louvres

Stack shut off damper

Cool air inlet

Store crane

Operating area

Product flask

Access plugs

Loading chute

Storage compartment

Typical section through product store

FIG 6 — PRODUCT STORE

Introducing autonomy to robotic manipulators in the nuclear industry

C L BODDY, BSc, PhD, AMIMechE
Advanced Robotics Research Limited, Salford
A W WEBSTER, BTech
British Nuclear Fuels plc, Sellafield, Seascale, Cumbria

SYNOPSIS The National Advanced Robotics Research Centre was set up in 1988 to provide a forum for the development and transfer to industry of the technology of Advanced Robotics. In the area of robot manipulators, research has been carried out into increasing the low-level autonomy of such devices e.g. reactive collision avoidance, gross base disturbance rejection. This groundwork has proven the feasibility of using advanced control concepts in robotic manipulators, and, indeed, indicated new areas of robot kinematic design which can now be successfully exploited. Within the newly defined BNFL Integrated Robotics Programme a number of joint projects have been defined to demonstrate this technology in realistic environments, including the use of advanced interactive computer simulation and kinematically redundant manipulators.

1 INTRODUCTION

British Nuclear Fuels plc (BNFL) has 15000 employees situated at five sites in the UK and its business activities span all aspects of the nuclear fuel cycle. In simple terms the activities include the enrichment of natural uranium, fuel fabrication, operating nuclear power generating stations, transporting, reprocessing and recycling spent nuclear fuel and management of the arising wastes. In recent years BNFL has seen an increased emphasis being placed on two aspects of its operations, namely safety and efficiency. Without a perceived high level of safety neither the regulatory authorities nor the public at large will underwrite the continued operation of nuclear plants. A high and increasing level of efficiency in the company's operations is also needed to enable it to withstand economic comparisons with other forms of energy production.

As part of BNFL's continuing policy of improving efficiency and safety, several robotic manipulator systems have been developed and introduced for various handling operations on three sites. Standard industrial robots are used in applications at Springfields and Capenhurst for fuel fabrication and size reduction of redundant plant respectively. At Sellafield some special purpose systems have been developed for handling tasks in the reprocessing and waste plants. An example called RODMAN is illustrated in Figure 1 which is designed to handle irradiated fuel rods within a small shielded containment vessel. This current range of robotic manipulator systems are all operating in well-structured areas of the operating plants and are successfully reducing operator radiation dose uptake and improving plant efficiency. In general they are operated in teach and repeat mode wherever possible, but many situations at Sellafield dictate teleoperation is used (i.e. man in the loop). Whilst the robotic manipulators are being used in this teleoperator mode, the principal feedback to the operator is usually from CCTV cameras.

Whilst there will continue to be applications within BNFL for robotic manipulators in well structured environments there is a growing awareness that the future remote handling challenges associated with operating and upgrading existing plants and decommissioning redundant facilities will be in less well-structured, ever changing environments, and ones which include a greater number of potential obstructions. BNFL's experience to date in developing manipulator systems to do engineering work in such environments (1) has demonstrated a requirement for a reduction in the amount of manual supervision of the remote handling task. Manual supervision is commonly required to ensure that no part of a robotic manipulator or the tool/object it may be carrying fouls any part of the plant as well as making sure the prime task is performed. The amount of man-effort required to interpret CCTV pictures and to teleoperate the equipment is the main factor governing the duration of the work. Increasing the amount of sensory data fed back to the operator will only exacerbate the problem. What is needed is a degree of autonomy where the supervisory computer receives and analyses the sensory data and makes decisions for the operator.

BNFL recognise that where new technologies are involved, work must be performed to demonstrate the technology in advance of a project application. Following on from this BNFL realised that no one company could afford to develop all the technology areas required to move present day robotics technology forward to what is now well known as Advanced Robotics. Hence BNFL's support for and collaboration with Advanced Robotics Research Limited (ARRL).

ARRL was founded in June 1988 to manage a new National Advanced Robotics Research Centre established at the initiative of the Department of Trade and Industry (DTI). The National Centre, which is located on the campus of the University of Salford, Greater Manchester, was started in collaboration with Salford University and a number of leading UK industrial organisations who, as shareholders, contribute research personnel and other resources to complement the start-up funding provided by the DTI. The principal research activities at ARRL are focused on the generic and enabling technologies of Advanced Robotics and their integration to achieve enhanced robotic functionality for a wide range of applications. The activities currently underway include research in the areas of task planning, vision systems, mobility, man-machine interfaces

(MMI), virtual reality, manipulator trajectory planning, and manipulator servo-control. The carefully planned research programme, being industrially driven, is structured towards achieving deliverable, robust, and commercially advantageous technology.

BNFL's research and development work in the field of robotics and telerobotics is being coordinated into what has been named the Integrated Robotics Programme (IRP). This programme draws together all previously separately controlled work in this field and provides a coherent structured route for the transfer and assimilation of the advanced technologies, including those developed at ARRL. The IRP comprises a series of Focused Application Demonstrator (FAD) and Research Demonstrator (RD) projects in which directed research and development will be performed to provide and demonstrate potential solutions to long term problems involving remote handling and robotics requirements, and to increase the level of expertise within the company in this important area.

Areas presently covered by the principal FAD's include Mobile Vehicles, Advanced Teleoperators, and Special Purpose Robotic Manipulators. Tasks are targeted at a range of timescales, from near term applications of "state of the art" mobiles in a drum store configuration to longer term assessment of uses for real-time three dimensional computer generated imagery. This paper discusses some of the research at ARRL in the area of robust manipulator trajectory planning, its influence on manipulator design, and the programme of work currently underway with BNFL to transfer this technology into the nuclear industry

2 ARRL ROBOT CONTROL ARCHITECTURE

ARRL is currently pursuing a wide ranging coordinated research programme based upon a standard functional architecture for advanced robotic systems (2). This incorporates a hierarchical control structure, enabling high level global planning to proceed with only an approximate representation of the external world. The resulting approximate plan can then be passed down the hierarchy for refinement and monitoring by sub-systems which handle local operational details. Only the lower levels of the hierarchy need to be intimately connected with the real-time control of input and output systems.

The functional architecture incorporates a 3 layer hierarchy, shown in Figure 2, which is based upon the principle of "increasing precision with decreasing sophistication" (3). Within this structure the highest level corresponds to the specification and task planning functions of the system, while the lower two correspond to the tactical and executive layers (trajectory planning and servo levels) respectively, of control. This structure possesses the advantage that each functional level possesses a clear and defined purpose, and has a clear interface to the next layer of the hierarchy.

The Strategic layer provides high-level autonomous function by formulating a strategy to achieve a set goal. High level operator goal/task commands will be decomposed into a series of achievable strategies, which will then be further developed into a series of specific device-related task commands by reference to the current static state of the system's world model. This level of control is aimed at replacing a significant proportion of the human input: a practice that is unlikely to be accepted in the nuclear industry until such systems have well proven track records. However, it is equally admissible for a human operator to replace this level of control using the standard tactical layer interface.

The Tactical layer transforms the sparse task commands into detailed trajectory commands for transmission down to the executive layer controller; i.e. it's central task is to carry out trajectory planning. Commonly, this is performed by converting the "world" task commands, in cartesian coordinates, into "joint" (or actuator) coordinates and applying polynomial profiling to generate smooth motion.

It is at the Tactical level of control that the low-level autonomous behaviour discussed in this paper can be introduced. Such behaviour is characterised by the utilisation of *a priori* information (for example, full or partial workspace representations) and of external sensory information (for example; object proximities, gross base disturbances, grasp mis-alignment) to make decisions and issue control instructions in a "transparent" fashion (relative) to the operator. The manipulator control system can thereby autonomously apply corrections (for example; potential collisions are avoided, large tracking errors are reduced, grasps are re-aligned) without reference to the operator. The overall aim of such autonomy is to reduce the amount of manual input an operator is required to provide.

The Executive layer performs the function of converting trajectory commands into actuator response. This is generally performed using closed-loop control with feedback from the internal joint sensors (or force sensors) to command the motors.

3 LOW-LEVEL AUTONOMY FOR MANIPULATORS

3.1 Real-time trajectory planning

Non real-time methods of trajectory planning, which are the norm in conventional robotics, become invalid when dealing with advanced robotic concepts such as working in semi-structured or unstructured environments. The generated strategic plan for a manipulator task, whether supplied by an operator or a strategic control layer, will always contain finite errors; for example due to (a) inaccuracies in the "world" model due to a dynamically varying environment, general modelling approximations, or significant "blind" zones, (b) approximations of manipulator dimensions and kinematics, (c) computational errors from data processing. To successfully cope with such situations it is necessary to dynamically compensate for such errors whilst the manipulator is actually performing a specified task i.e. the manipulator possesses a level of autonomy.

The functionality of the tactical layer control can thus be seen to include the need to modify, at a local level, elements of the strategic plan; furthermore, to do so in real-time as the manipulator interacts with the environment. This functionality can only be facilitated by the existence of a central real-time trajectory planning algorithm to update the ongoing trajectory plan. Using this as the basis for a control system design facilitates extendable functionality in a modular fashion (4).

3.2 ARRL manipulator research demonstrators

The use of low-level autonomy within manipulator trajectory planning systems can have many applications. Two examples of such control systems have been developed at ARRL, as part of a coordinated research program, through the conceptual development stage and into working demonstrator systems. Both have been developed in the framework of a modular control system design.

The first manipulator research demonstrator is a low-level collision avoidance control system where proximity sensing devices are mounted on the "skin" of the robot, around the wrist area, to detect potential collisions (see Figure 3). If detected obstacles represent a possible collision condition, where this criterion is evaluated using the sensor velocity towards the object and the distance (5), then the collision avoidance control system generates a notional repulsion vector which is determined such that it "drives" the end-effector away from all possible collisions. This repulsion vector is then passed to the manipulator trajectory planner, in the form of an intermediate task point, which then interrupts the existing plan passed to the control system by the operator.

The trajectory planner is essentially the core module within a robot tactical layer controller. Given either a single goal point or a sequence of goal points with specified motion types between points (world or tool space) it will generate the necessary actuator (joint space) motions to achieve the desired tip motion. This task is conventionally performed off-line in current industrial manipulators, particularly in the case of straight-line or complex cartesian motion where a kinematic inversion must occur every cycle of the controller. However, given the requirement of the low-level collision avoidance control system to incorporate repulsion motions in real-time, the trajectory planner had to be designed to run in real-time (as discussed in section 3.1).

The behaviour of the manipulator is essentially reactive, since the control system makes no attempt to model its environment or store past object proximity data, and relies on a series of repulsive motions to circumnavigate larger objects. It has clear applications in areas where manipulators are required to operate in an environment with obstacles with (a) the ability to work with some autonomy when no model is available, and (b) the ability to work successfully with low resolution world models, or teleoperators with incomplete knowledge of the environment. This type of avoidance control can be seen as a "local" safeguard.

The second relevant manipulator research demonstrator project at ARRL is the gross base disturbance rejection project (6). The aim of the project is to stabilise the end-effector of a robot manipulator which has unknown position disturbances at its base (see Figure 4). External position and orientation sensing systems are used as a feedback to determine required pose (position and orientation) corrections of the manipulator. From this definition it is clear that the use of a real-time pose planner is required, analogous to the trajectory planner used in the reactive collision avoidance research demonstrator.

The disturbance to the robot base is simulated by a hydraulically actuated 3 degree of freedom (dof) platform which has heave, pitch, and roll motions available, thereby providing a large variety of possible test disturbances. A laser tracking system is one of the main external sensor systems used for disturbance rejection.

This technology has application in areas where manipulators without rigid base mountings are used but which are required to maintain fixed end-effector positions relative to the "world" coordinates: typically this could be on an underwater vehicle or mobile vehicle on rough terrain, or even a space application.

3.3 Kinematic redundancy of manipulators

The number of degrees of freedom available at the end of a serial kinematic chain, N_{dof}, (e.g. conventional robot manipulator) is equal to the number of joints, assuming a degenerate configuration is not chosen. More formally

$$N_{dof} = 6(n-1) - \sum_{i=1}^{j} (6-f_i) \qquad (1)$$

where n is the number of links in the chain (including the "ground" link) and j is the number of kinematic pairs (or joints) with f_i associated dof. Almost all conventional manipulators have six joints (or less) this being the minimum required for full 6 dof motion at the end-effector. Manipulators with more than 6 joints are said to be kinematically redundant, and are relatively uncommon. Typical examples are the GEC Advanced Slave Manipulator and the Robotics Research Corporation K series manipulators.

The advantages of such kinematic configurations are clearly many-fold. The available workspace of the manipulator is significantly increased since on conventional manipulators much of the internal workspace is physically unreachable due to mechanical joint limits. Thus the available dexterity at any position in the workspace is also increased, i.e. the range of continuous movement available at the end-effector. Over a work volume this may imply, for example, that one dextrous manipulator may be capable of replacing two or three conventional manipulators in a work-cell. A property related to this is that a redundant manipulator can reach spaces with tricky access yet retain full 6 dof motion at the end-effector. Lastly, such manipulators have the ability to perform sub-tasks such as resting the manipulator against solid surfaces to reduce end-effector compliance during force operations, collision avoidance whilst maintaining desired end-effector motions, joint torque optimization, or avoidance of singularities (a major problem with conventional manipulators).

It is therefore, perhaps, surprising that more such manipulators are not commercially available given the broad range of tasks that they can perform. The fundamental problem with controlling the motions of such manipulators lies in the infinite number of solutions available for the configuration of the manipulator for any given desired position and orientation of the end-effector. In other words, the crucial problem of translating desired "world" trajectories into "joint" trajectories is under-constrained. But, of course, it is this very lack of constraints that gives kinematically redundant manipulators their desirable properties. The RRC range of manipulators, mentioned earlier, is close to an anthropomorphic configuration, and can be controlled by using the "elbow roll" angle as the choice for an additional constraint (7). The GEC Advanced Slave Manipulator requires that the operator lock one of the joints during controlled tip motion, thereby reducing the manipulator trajectory control to the conventional 6 dof problem.

Modern kinematic control techniques have now provided solutions to the more general problem. Possibly the most well known solution is that of "pseudo-inverse" control which uses the Moore-Penrose pseudo-inverse of the manipulator Jacobian to generate the required joint velocities to perform a given task (8). (Note a conventional matrix inverse does not exist since the Jacobian is by definition non-square). However, this solution has associated problems including the fact that it does not avoid, or control, singularities. A "damped least squares" pseudo-inverse based on a modified solution was proposed in (9), which circumvents this problem.

Such solutions can also be augmented to include sub-tasks, by projecting "joint" space sub-tasks onto the null space of the manipulator Jacobian (10). The null space of the manipulator Jacobian represents the space of instantaneous joint motions which will result in zero motion of the end-effector and is a crucial concept in the control of redundant structures.

By incorporating proximity sensors over large areas on such manipulators, the methods outlined earlier in section 3.2 for the manipulator end-effector collision avoidance control system, can be used to generate elemental repulsions (4) and thus suitable "joint" space sub-tasks may be constructed to provide whole manipulator collision avoidance (11). This particular aim is the subject of a joint project between ARRL and BNFL under the auspice of the FAD programme, and it is possible that a kinematically redundant manipulator will be purchased in the near future for the development of such a system.

4 IMPLEMENTING THE TECHNOLOGY

It is clear that BNFL's IRP will not achieve overnight success in providing an autonomous robotic manipulator which can be used in a plant tomorrow or even in one years time. The programme is an incremental one in which gradually more advanced technologies will be introduced over a number of years. Fortunately, the successes achieved in robotic manipulator projects over the last 5 years have done much to generate a positive attitude to the introduction of new technologies with the plant designers and operators. It is intended to foster this attitude through their participation in the FAD and RD programme.

The robotic manipulator system described earlier and shown in Figure 1 has already done a great deal to gain acceptance for the concept of autonomy within BNFL. RODMAN is equipped with a simple collision avoidance system which freezes the motion of the manipulator whenever any part of it comes within 50mm of the

environment in which it works. The operator must then select a boundary override mode and take responsibility for preventing collisions. When the manipulator returns to the safe envelope the motion is frozen once more to prompt the operator to turn the boundary override mode off. Using this system the operators can drive the manipulator with confidence in open loop and only need to concentrate on the performance of the task rather than worrying about potentially damaging collisions. However, this system relies on the well defined, unchanging geometry of the area in which the manipulator works.

In the first year of the IRP it is intended to do work in two specific areas to move towards autonomy. The first is in the area of computer simulation and modelling. Here, one workpackage is intended to demonstrate that the simulation can be driven by the outputs from a robotic manipulator system working in a complex structured/semi-structured environment. Figure 5 shows an early computer generated model for this example which was used to validate the design on the equipment. The aim is to use the simulation to warn the operator of potential collisions and to provide him with simulated views of the equipment (to help him perform the task) which are unavailable in practice. This may seem to be little more than has been achieved already by RODMAN, but it will involve a much more complex environment and the 7 degree of freedom GEC manipulator. It should be possible to determine whether the sources of error noted in Section 3.1 are significant.

The second area of work in the IRP which will pave the way towards autonomy is whole arm collision sensing and collision avoidance applied to kinematically redundant arms, as described in Section 3.3. Once the initial work is completed in these areas it should be possible to demonstrate a robotic manipulator system which can be teleoperated easily, is self protecting, can maintain the desired tool trajectory and orientation whilst avoiding collisions and can present a simulation view of the task which holds sufficient information for the operator to act upon.

In future years of the IRP it is intended to incorporate advances in other areas such as task planning, mobility, world modelling, virtual reality and other improved man machine interface technologies.

5.0 CONCLUSIONS

True autonomous behaviour of robotic devices is still a considerable time away, but many of the key underlying technologies are currently being researched at ARRL. The immense complexity involved coupled with the non-deterministic behaviour of such systems promises many headaches for those involved in safety issues, even in less safety critical areas than the nuclear industry.

In the shorter term, the technologies described in this paper can be used both to introduce improved control systems that allow reduced operator load, increased cycle time, and improved safety mechanisms, and to introduce the use of complex kinematic designs, such as kinematically redundant manipulators, that will broaden the task range of manipulators. This technology coupled with the improvements offered by improved human-robot interfaces (12) will be that which is seen in the nuclear industry over the next decade.

BNFL's new Integrated Robotics Programme and its links with ARRL should not only provide the technological advances necessary to carry the company into the next century but also to provide, within the work force, the enhanced expertise necessary to sustain them.

REFERENCES

(1) Jones, E.L. and Webster, A.W., "Remote diversion of a highly active process pipeline", IBC Conference on Remote Technology for the Nuclear Industry, London, 1990.

(2) Pegman, G.J. and Gray, J.O., "The National Advanced Robotics Research Centre: a research programme overview", Proceedings of the Institute of Mathematics Applications Conference on Robotics, Loughborough, 1989.

(3) Saridis G.N. "Towards the realisation of intelligent controls", IEEE Proc. 67, [8], 1979.

(4) Boddy, C.L. and Stobart, A. "Application of a hierarchical control structure in the implementation of a manipulator real-time collision avoidance system", 5th International Conference on Advanced Robotics, Pisa, Italy, June 1991.

(5) Boddy, C. L. "Implementation of a real-time trajectory planner incorporating end-effector collision avoidance for a manipulator arm", 2nd Int. Workshop on Advances in Robot Kinematics, Linz, Austria, 1990.

(6) Manganas, A.M. "The stabilisation of manipulator tip motions in response to low frequency, high magnitude manipulator base disturbances", ARRL Internal Report, No. RD4/CS-2, 1990.

(7) Seraji, H., Long, M. and Lee, T. "Configuration control of 7 dof arms", Proc. IEEE Int. Conf. Robotics and Automation, Sacramento, California, 1991.

(8) Whitney, D.E. "Resolved rate motion control of manipulators and human prostheses", IEEE Trans. Man-Machine Systems, 10, pp47-53, 1969.

(9) Nakamura, Y. and Hanafusa, H. "Inverse kinematic solutions with singularity robustness for robot manipulator control", Trans. ASME J. Dynamic Systems, Measurement, and Control, 108, 1986.

(10) Nakamura, Y. "Advanced Robotics, Redundancy and Optimization", Addison-Wesley Publishing, Reading, Massachusetts, 1991.

(11) Espiau, B. and Boulic, R. "Collision avoidance for redundant robots with proximity sensors", 3rd Int. Symposium Robotics Research, MIT Press, 1986.

(12) Stone R J "The best of both worlds: a combined virtual-real human-computer interface for telepresence and remote driving", International Symposium on Advanced Robotic Technology, Tokyo, Japan, 1991.

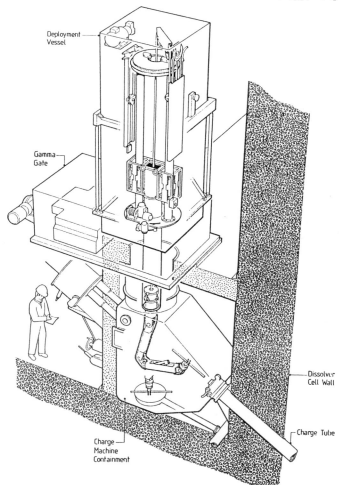

Figure 1: "RODMAN" Charge Machine Manipulator Handling a Fuel Rod

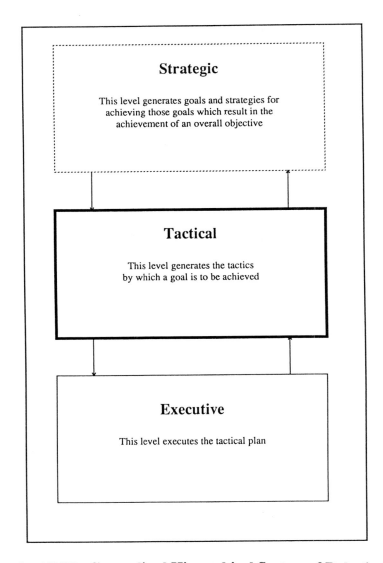

Figure 2 - ARRL Generalised Hierarchical System of Robotic Control

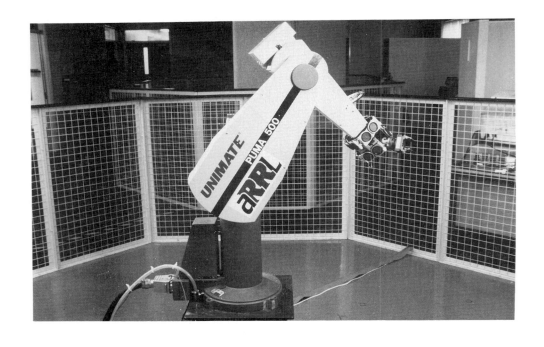

Figure 3: Manipulator End-Effector Collision Avoidance Demonstrator

Figure 4: Manipulator Gross Base Disturbance Rejection Demonstrator

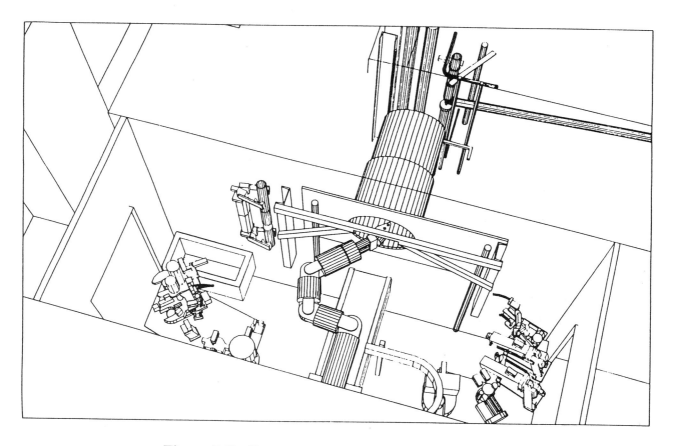

Figure 5: Raffinate Pipe Diversion Workspace Layout

C431/060

Remote handling systems, options and designs for handling radioactive waste packages

A M McCALL, BEng, AMIMechE
UK Nirex Limited, Harwell, Didcot, Oxfordshire
G K WILSON, BSc, CEng, MIMechE
BNFL Engineering, Risley, Warrington, Cheshire

UK Nirex Ltd (Nirex) has the task of providing a facility for the final disposal of radioactive waste from the nuclear industry activities within the UK. Nirex is currently developing conceptual designs for this facility for the disposal of radioactive waste deep underground. This paper provides a general overview of the mechanical handling issues which have arisen as the design progresses. As the design is undergoing rationalisation based on this work, it is intended that the presentation of this paper in November 1991 will be expanded on by means of diagrammatic representations of the approach and conceptual designs described.

1. INTRODUCTION

This paper presents the conclusions regarding mechanical handling aspects of the design studies undertaken by Nirex during 1990 for the surface facilities only. The aim of the studies was to develop the design of the surface facilities required for the repository for radioactive waste taking into account the two potential sites favoured by Nirex, Sellafield and Dounreay. The paper starts by describing the main objectives and aims of the design studies, and then goes on to describe the packages to be hanlded and the philosophy developed for preparing such packages for disposal. Each of the handling lines is briefly described

The objective of the studies was to examine in detail particular areas of the conceptual building, civil and mechanical design of the surface facilities. In addition the studies examined the potential advantages which might be available through the use of technology and practices not normally used in the nuclear and mining industries.

In particular the specific issues to be addressed in siting the facilities at Sellafield or Dounreay were examined. The design of the facilities took into account the latest information on waste packages and the preparation requirements for those packages.

1.1 Approach

The studies were split into five main areas of work:

 Design Definition
 Mechanical Design
 Ventilation
 Civil Design
 Architectural Overview

The **Design Definition** was mainly concerned with establishing the interface between the waste to be delivered to the surface facilities and its preparation into a form suitable for placement deep underground.

The object of the **Mechanical Design** was to examine the mechanical handling options for each aspect of the waste handling within the repository facilities taking due account of the requirement for an interface between the surface and underground mechanical handling systems.

The **Ventilation study** examined the ventilation system requirements within the surface facilities and then considered the effect of the discharges from the repository on the local environment.

The object of the **Civil Design** was to assess the various site and building layouts.

The object of the **Architectural Overview** was to examine the entire surface facility and to assess its visual impact on the local environments at Dounreay and Sellafield.

The study was not a detailed design but a preliminary investigation of the issues which will affect repository design when a single site has been selected. In order to identify these issues specific design solutions to particular problems were generated, accepting that the availability of more detailed information later in the programme may require a re-evaluation of the preferred options presented in this paper. Nirex is continuing to rationalise and develop the conceptual designs arising from this study and it is anticipated that the philosophy will change as design development progresses.

The study did not make any evaluation of the benefits of the Sellafield site over the Dounreay site or vice-versa.

1.2 Waste Arrival

The purpose of the repository is to provide a safe and permanent disposal site for intermediate and low level nuclear waste. This waste will arrive at the surface facilities by road and rail and in various standardized packages. Rail deliveries will be by dedicated trains. Locomotives will be owned and operated by British Rail. On arrival at the site boundary wagons will be shunted into sidings where they can be picked up by a Nirex locomotive for transfer into the site. Road deliveries will be made to the site and trailers parked in the road storage area prior to being picked up by a Nirex unit and taken for unloading.

Sellafield and Dounreay data were considered separately, due to the fact that over 60% of ILW originates at Sellafield. Therefore at Sellafield it was assumed that the waste receipt building would have a link with BNFL plant to enable direct transfer of waste packages where possible. In the case where waste packages arise on an intermittent basis, for instance decommissioned fuel flasks, referred to as "Large Items", the consignors may choose to stockpile the packages until a full trainload of wagons can be assembled.

It is Nirex's current intention that should the repository facilities encounter problems such as industrial action then all waste packages currently in transit should be accommodated within the site. The site is therefore provided with large sidings. Unloading of packages from rail transport is envisaged as similar to the handling in freight container yards.

1.3 Waste Handling and Treatment

Low level and Intermediate level radioactive waste in the UK takes various forms. It is not the objective of the surface facilities to provide any waste treatment processes except ensure that the waste is in a form suitable for disposal underground.

Hence a major feature of the mechanical handling study was the rationalisation of the various waste packages arriving at the facility. The various packages are shown in Figures 1, 2, and 3. The majority of the waste packages may require inspection and monitoring to confirm that the waste is indeed in a suitable form for disposal. However two categories of the waste package were assumed to require some treatment in order to establish design options.

1. Firstly the large Low Level Waste boxes (fig. 1) may be required to be filled with cementicious grout. This could be carried out in the waste receipt building in the Low Level Waste grouting line, an area of the plant specifically designed for this purpose.

2. The second category of waste which may require treatment is intermediate level waste (fig. 2) which requires temporary biological shielding during transport and transfer operations. These packages are typically 500 litre drums and large drums and boxes. These packages arrive at the waste receipt building in the Reusable Shielded Transport Containers (RSTC's). The RSTC will hold either four 500 litre drums or one of the larger items. Ideally, to reduce double handling, the packages would be transferred directly to the vault in the RSTC and only unloaded on arrival at the vault. However in the absence of data on the underground system at the start of these studies, it was assumed for the purpose of this study that this would complicate the operations required underground to an unacceptable degree. The packages were therefore assumed to be inspected and monitored at the surface. They would then be loaded into a large capacity container providing temporary shielding during transfer underground this is known as the Vault Shield Box (VSB). The Vault Shield Box has a capacity of 8 off 500 litre drums or 2 larger ILW packages.

The surface study grouped the waste package according to the waste handling philosophy to be adopted in the Waste Receipt Building, as follows:-

Low Level waste packages which may require grouting.

Unshielded Intermediate Level waste packages which need to be remotely handled behind biological shielding.

Large miscellaneous self shielded packages which can be directly handled and which do not require grouting.

Reusable packages consisting of the reusable shielded transport containers and Nirex transport stillage. These items are not destined for disposal in the repository but are used for the transportation of the unshielded intermediate level waste packages discussed above.

The rationalisation of the packages revealed that three waste preparation lines would be required.

a) Low Level Waste grouting line.
b) Intermediate Level Waste line (biologically shielded).
c) Large Items line (fuel flasks).

This formed the basic philosophy of the waste receipt handling area.

There are a number of items which are not listed as Nirex standard packages but will be required for the handling of the waste packages described above, briefly they are described as follows:

Disposable stillages to accept four 500 litre drums following removal from the reusable shielded transport container. Drums which arrive in the RSTCs will not be carried in a disposable stillage but in a purpose designed transport stillage. The transport stillage will have a number of features which make it too expensive for disposal, a disposable stillage will therefore be required. Stillages are required to facilitate handling throughout the waste preparation and handling sequence and to enable the drums to be stacked within the repository caverns.

Special handling frames will be required for the decommissioned fuel transport flasks to facilitate handling and transportation throughout the disposal sequence. The frames are essential if these flasks are to be handled and stacked remotely underground.

2. LOW LEVEL WASTE GROUTING LINE

A facility to grout LLW boxes may be provided within the Waste Receipt Building and was included in the surface design studies. The purpose of grouting being to increase the crush strength and hence allow LLW boxed to be stacked higher than the 6 high required of the IAEA transport regulations. This will enable a cost comparison to be obtained between addressing this problem in the surface buildings or in the underground vaults by the use of structural supports.

When large LLW boxes are transferred into the grouting line they will first be moved to the grouting position, and then moved to a storage area to stand whilst the grout cures. When curing is complete the box will be taken out of the line to an inspection area. The type of equipment proposed to carry out these movements was a specially designed rail mounted bogie.

When the boxes have been grouted, cured and inspected they will be ready for transfer to the Vault Transfer Vehicle (VTV). An overhead travelling crane was selected for this operation. The crane will be supported on the building structural steelwork and will be able to access the despatch areas of all three grouting lines.

3. INTERMEDIATE LEVEL WASTE HANDLING LINE

The ILW line provides a means of transferring the 500 litre drums from the 4 drum RSTC to a Vault Shield Box with a capacity of 8 drums. This will reduce the number of transfer operations to the repository and simplify the handling operations underground. The line also provides a means for quality assurance checking of the various packages prior to disposal. These checks will include visual inspection, package identification, and monitoring for surface contamination.

To accommodate any items that fail inspection an 'Off Spec' line can be provided. In this line it was envisaged that reject items will be overpacked in large self shielded boxes and grouted into place in order to provide a clean, mechanically sound package for transfer to the repository.

3.1 Description of Waste Handling Operations

The operations described below commence with an RSTC having been inspected and monitored in the inspection hall in readiness for transferring into the ILW line.

Using an overhead crane the RSTC will be transferred onto a waiting bogie. If the RSTC is known to contain 500 litre drums located in the non-disposable stillage then a disposable stillage will also be placed onto the bogie before it enters the ILW Line.

The bogie carrying the RSTC and stillage will then be moved through a heavy shield door into a biologically shielded area for subsequent operations.

The RSTC lid will be removed at the first station and the bogie will move forward until its contents are directly below the lifting station. The RSTC contents will be lifted out and the bogie will move back so that the stillage is directly below the waste packages. The waste will then be lowered into the stillage. The disposable stillage is then lowered onto another bogie which moves through a shield door into the main ILW cell. Items which do not require the disposable stillage will be placed directly onto the second bogie for transfer into the main cell.

When the RSTC bogie has been withdrawn from the unloading station the lid removal machine will replace the lid allowing the bogie to withdraw to the lid unbolting/bolting station where the lid will be bolted into place. The bogie will then take the RSTC out of the shielded area where it can be lifted into the inspection area by the overhead crane prior to inspection, monitoring and return to the waste consignor.

In the ILW cell waste packages will be moved using an overhead crane. Three inspection stations will be provided, each equipped with a rotating table placed in front of a lead glass window where packages can be visually inspected. A package recording system will automatically record the type of package and its identification number whilst the package is being visually inspected. A swabbing facility will be located adjacent to the rotating table which will swab a portion of the package to check for surface contamination.

At Dounreay all ILW waste packages will arrive in RSTCs, however at Sellafield approximately 60% of the ILW packages will arrive from adjacent BNFL plants. This could will be via a tunnel which would terminate under the main ILW cell. This would allow the overhead crane to lift the drums (which will be already packed in a disposable stillage) directly from the tunnel to one of the inspection positions.

Packages which are accepted by the inspection system for disposal will be transferred using the second overhead crane to the buffer storage area. When the export route is available the packages will be moved using the second crane through a heavy shield door into the dispatch area where they will be lowered into the Vault Shield Box.

Packages which are not acceptable for direct disposal will be transferred into the adjoining "off-spec" grouting area. Here they will be grouted into large self-shielded boxes before being transferred to the dispatch area via a shield door.

3.2 Maintenance Workshop

The maintenance workshop is situated in line with the ILW and "off-spec" dispatch areas. This workshop is used for maintaining the ILW and off-spec overhead cranes, in-cell equipment and transfer bogies. The equipment is all modular in form and the cell can be fully dismantled to facilitate maintenance and good housekeeping for the 50 years of operation.

4 LARGE ITEMS LINE

The purpose of the large items line is to provide a facility for preparing miscellaneous radioactive waste packages usually from decommissioning operations for disposal in the underground repository.

Such items include decommissioned fuel transport flasks, overpacked Multi-Element Bottles etc. These items have no common handling features and the primary function of the Large Items Line is to place each item into a lifting/stacking frame. Though each frame will be designed specifically for a particular waste package it can have standardised lifting and stacking features which will allow these miscellaneous items to be handled and stacked remotely underground.

In addition this line has a facility for the routine maintenance of the fleet of RSTCs used in the transfer of ILW to the repository. This maintenance facility and the general area within the large Items Line may also be used for repairing and decontaminating damaged RSTCs and large LLW boxes from the low level waste lines. This line also offers a service facility for Vault Transfer Vehicles and Vault Shield Boxes.

5 WASTE EXPORT AND SHAFT HANDLING FACILITIES

The transfer of waste packages from the Waste Receipt Building to the Waste Shaft is carried out by means of a number of independently controlled rail vehicles which will transport packages from the Waste Receipt Building (WRB) to the underground vault.

At both Sellafield and Dounreay the rail link between the WRB and the Waste Shaft was assumed to be underground, a sub-surface tunnel was selected as the preferred option.

Facilities are provided at the despatch end of each of the waste lines in the WRB in order that waste packages can be lowered directly onto the vault transfer vehicles which may contain a Vault Shield Box as appropriate.

The Vault Transfer Vehicle (VTV) will have a capacity to allow it to transport any of the waste packages. Location and package securing features will be designed to be adaptable to accommodate the various package designs. The vehicles will be electrically powered. It is envisaged that each vehicle will be uniquely identifiable by the control systems both above and below ground in order to control the routing of individual waste packages.

Intermediate level waste packages will require biological shielding during transport to the underground repository. The Vault Shield Box provides this shielding. To minimise the number of vault transfer operations each VSB will carry two ILW packages, or eight drums. In line with the remote handling philosophy specially designed recovery equipment will be used to retrieve the VTV in the event of equipment failure.

To simplify mechanical handling operations throughout the waste package was assumed to remain on the vault transfer vehicle during lowering.

UK Nirex Ltd have also completed conceptual design work on the underground aspects of the repository. This has enabled Nirex to examine the design issues and solutions with a view to rationalising the design of the repository as a whole.

6 CONCLUSION

The surface and underground design studies carried out to date have shown that even though waste consignors have a need to use a variety of packages for radioactive waste a mechanical handling philosophy commensurate with remote operation can be foreseen.

The design studies to date are at the conceptual stage and have enabled specific design solutions to be generated. However, further evaluation of the concept is a continuing process and further change and development is ongoing.

The facilities to be provided by Nirex are simply to receive waste packages for final disposal. The task is mainly based on mechanical handling and as such aspects from the freight container industry, the nuclear industry and the mining industry provide an engineering solution to the disposal of solid radioactive waste. © UK Nirex Limited 1991

7 REFERENCES

1. BARLOW S. et al. The Interaction between the Design of Waste Containers for Low Level Radioactive Waste and the Design of a Deep Repository. This Conference, November 1991.

2. IAEA. Regulations for the Safe Transport of Radioactive Material (1985 edition as amended 1990) IAEA, Vienna 1990, Safety Series 6.

3. HOOPER A. ISGAR P. Repository Design Strategy. Civil Engineering in the Nuclear Industry (ICE) March 1991.

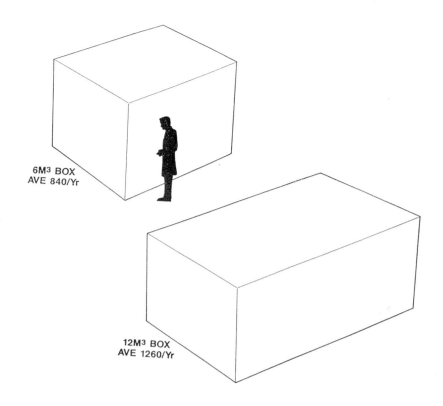

Fig 1 Low level waste packages

6M³ BOX
AVE 840/Yr

12M³ BOX
AVE 1260/Yr

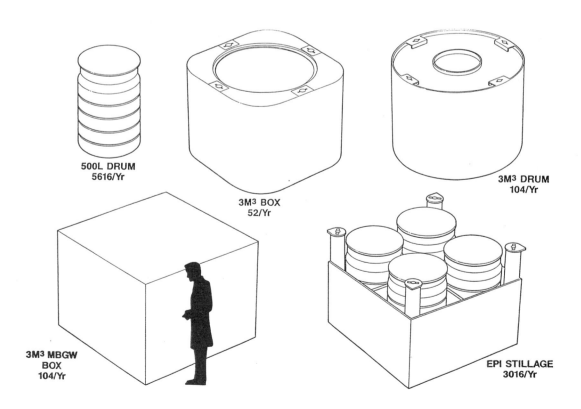

500L DRUM
5616/Yr

3M³ BOX
52/Yr

3M³ DRUM
104/Yr

3M³ MBGW
BOX
104/Yr

EPI STILLAGE
3016/Yr

Fig 2 Intermediate level waste packages

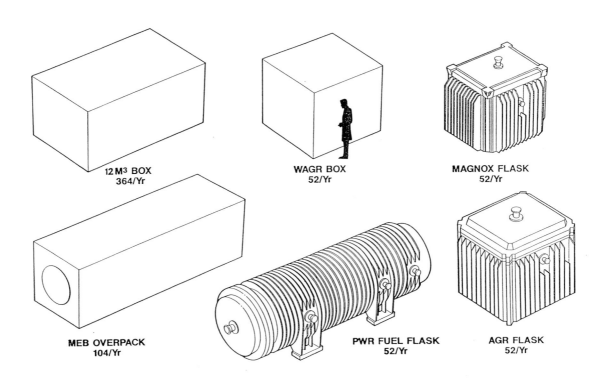

12 M³ BOX
364/Yr

WAGR BOX
52/Yr

MAGNOX FLASK
52/Yr

MEB OVERPACK
104/Yr

PWR FUEL FLASK
52/Yr

AGR FLASK
52/Yr

Fig 3 Large items

Shaft hoisting techniques

J NEWSUM, CEng, FIMechE and K W LOOMES, BSc, CEng, MIME
Qualter, Hall and Company Limited, Barnsley, Yorkshire

Shaft Hoisting Techniques

SYNOPSIS

SYNOPSIS   This is a description of shaft hoisting techniques, based on known and proven mineshaft technology used for a deep level underground repository for radioactive waste.

A depth between 500 to 100 metres would be required accessed by a minimum of two vertical shafts for safety considerations and to facilitate ventilation. We have assumed that the vertical shaft concept has been selected on merit from other engineering concepts and the 1000 metres depth is to suit the geological and hydrological conditions at the selected site.

The requirement is for the geological environment chosen to host the repository, to back up the engineering safeguards in preventing the migration of radionuclides back to the biosphere.

The packages to be transported underground would be in the weight range 20 to 110 tonnes. The largest volume being 6000 x 240 x 2750.

The waste packages have to remain monitorable and retrievable during the period prior to the closure of each repository vault.

The Civil Engineering design of the shaft wall has to be considered with reference to the determination of the shaft diameter. This takes into account the geological and hydrogeological conditions of the chosen site. This means that the designers of mechanical and structural equipment don't have a completely free hand in determining the diameter of the shaft.

In basic terms the plan area of the package determines the shaft diameter and the weight of the package determines the type and power of the winding system.

1 THE BACKGROUND INTRODUCTION

The repository is to store low level waste and intermediate level waste only. Regarding the high level waste, over the past 30 years only 1200 cubic metres have been generated in Britain and the present policy is to store HLW at Sellafield Reprocessing Plant for a period of 50 years.

At present roughly 2500 cubic metres of ILW are produced in Britain each year. In addition, roughly 40 000 cubic metres of ILW is currently in storage. About 26 000 cubic metres of low level waste are produced in Britain each year.

On average there will be either six train-loads or four trainloads and 33 lorry loads of radioactive waste arriving at the repository every week.

The radioactive waste transported to the repository by rail and road transport would be contained in boxes or drums.

The waste would be transported from the surface reception facilities to the underground storage vaults in shielding boxes which would later be returned to the surface facility.

The items to be transported down the vertical shaft in the cage would be the waste in its shielding box carried on an eight wheeled self powered vehicle with an integral braking system and operated by remote control. Assume the package we are handling is the following weight and size.

Maximum weight of package	65 tonnes
Weight of vehicle	20 tonnes
	85 tonnes

Size of package 6000 x 2750 x 2400
Size of Vault Transfer Vehicle 6400 x 2940 x 850

From the size and weight of the load to be carried by the cage a few simple calculations show that a shaft diameter in the order of 8000 is required for the 6400 x 2940 plan area vehicle and that a multi-rope tower mounted (see Fig I and Fig II) friction winder is required to handle to 85 tonnes load.

The cage would travel at a speed of three metres per second in the shaft to handle the volume of waste. This is a relatively slow speed when compared with mining practice.

1.1 Radiological Protection Zones

The calculated radiation dose rate outside the shield box is such that within the confines of the shaft top building, the vertical shaft and the underground tunnels, occupancy around the vault shield box is unacceptable.

Therefore the movement of the package from surface to underground vault must be achieved by remote control operations.

2 THE WINDING TOWER AND TYPE OF WINDING ENGINES

Winding towers have been constructed in all steel construction or concrete construction. The site location and location of the winding engine are contributory factors in the decision making process. Drum winders are known to be exclusively ground mounted. Friction winders are either ground or tower mounted. In general, if there is sufficient space at ground level two rope friction winders are often quicker to construct as ground mounted winders. However, with ground mounted friction winders the number of ropes is normally limited to two in order to avoid complications which are expected with a large number of head sheaves.

North American experience indicates tower mounting gives a definite cost and operating advantage particularly if more than one winder is involved.

Where severe winder climatic conditions are applicable, using slip formed concrete towers which completely enclose the winding system gives better protection than steel towers.

For the application being considered because of the number of winding ropes. Tower mounting of the multi-rope friction winder is the preferred choice.

2.1 Types Of Winders

DRUM WINDER with a parallel cylindrical drum, with or without balance rope. This is the simplest form of drum winder.

Large drum winders today are driven by low speed direct coupled electric motors flanged to the winder shaft without the need for an external bearing.

DOUBLE DRUM WINDER If winding is to be carried out from two or more levels in a shaft, it is convenient to use two cylindrical drums. Two equal cylindrical drums are used, one of which is keyed to the drum shaft the other operated through a clutch.

BLAIR WINDER This is a two rope, double drum winder used mainly in South Africa. One installation in the United Kingdom is at the Cleveland Potash mine in North Yorkshire.

FRICTION WINDERS See Fig V. Presently multi-rope friction winders in mines have up to ten winding ropes. Lifts in buildings are a type of friction winder.

To ensure that the winding ropes do not slip on the winder drum in all operating conditions, it is necessary to use balance ropes. The ratio of T1 to T2 (weight of full side to weight of empty side) should not exceed 1.6. In the case of shallow shafts with correspondingly low rope weights there is a low dead load and thus a low permissible payload.

Large friction winders have, like modern drum winders, direct coupled electric motors flanged to the main shaft in an overhung position. On the European continent some winders now combine the drum and winder so that the stator of the motor is constructed within the drum shell.

Today's shaft winding installations are so reliable that the previously used arguments for the division of the shaft into two winding systems (to continue operations with half the capacity in case of failure of one installation) are no longer viable. Therefore, modern winding installations comprise one single large system.

Modern shaft signalling, communications, and winder safety monitoring systems are now available using the winding ropes as the transmission path. This development can be used for the following:-

    Shaft Signals
    Shaft Speech Communications
    Slack Rope monitoring/Evaluation
    Conveyance Weighing
    Rope Slip Monitoring

2.4 Multi-Level Winding

The normal duty for most winders is to operate between fixed levels at the top and bottom of a shaft. There are, however, some shafts which have one or more insets at intermediate levels, to which service must be provided.

For occasional winds the conveyance may be stopped at an inset, the engineman winding to a special mark on the depth indicator.

2.5 Conveyance and Counterweight (See Fig IV)

This is a useful system for winding in a shaft requiring a frequent service to more than one level. It is also used where a conveyance of large plan area is required. The counterweight is made equivalent to the weight of the conveyance plus half the weight of the load.

2.6 Winder Power Required

In the case of the winder under consideration, the cage and counterweight system has 155 tonnes on the cage side and 112.5 tonnes on the counterweight side. Therefore the maximum out of balance load the winder has to deal with is 42.5 tonnes.

With a winding speed of three metres per second, the power of the winder motor would be in the order of 2000 kilowatts.

3   THE WINDING AND BALANCE ROPES ALSO ROPE STRETCH

To carry the loads involved would require 8-52 millimetres diameter locked coil winding ropes with 6-73 millimetres dia balance ropes.

Fig III shows typical construction of these ropes.

In the majority of mining communities such as United Kingdom, Germany and South Africa a philosophy of safety has evolved over many years of practical experience. This is to have a high factor of safety on the winding ropes, with regular inspections of the ropes and recapping every six months and new ropes every two years. In deep vertical shafts this is found to be the safest and most practical approach.

SAFETY GEAR The following is an extract from a Health and Safety publication.

Conveyance safety-gear has never been required by law in shafts in Great Britain or favoured as a means of protecting persons. It has been the practice to rely on the mechanical brake of the winding engine, the automatic contrivance and on the integrity of the winding rope and suspension gear.

This practice, combined with a high standard of inspection and maintenance has virtually eliminated failures in service.

WINDING ROPE DETAILS - 8.52 millimetres dia full locked coil.

    Weight 15 kilogrammes per metre.
    Total weight 1000 metres x 15 x 8 = 120 tonnes.
    Breaking load 233 tonnes per rope x 8 = 1864 tonnes.
    Factor of Safety 6.73 to 1.

BALANCE ROPE DETAILS - 6-73 millimetre dia

    Weight 20.4 kilograms per metre.
    Total weight 1000 x 20.4 x 6 = 122 tonnes.
    Breaking load 198 tonnes per rope.

3.2 Rope Stretch

Because the ropes are constructed from steel wires they stretch under load. At 1000 metres deep with the 85 tonnes load in the cage the ropes will stretch about 520 millimetres.

If the rails in the cage are brought level with the rails at the fixed landing at the shaft bottom, the self propelled vault transfer vehicle can be driven out of the cage.

When the first set of wheels leave the cage 42.5 tonnes weight is removed from the winding ropes, the tension in the winding ropes will reduce, the ropes will relax and the cage floor will lift 260 millimetres such that the rails are no longer in line.

This problem is common in mining practice and to deal with it there are, in the main, two types of equipment.

(1) To have a drawbridge (hinged platform) fitted into the fixed landing at each end of the cage.

     When the cage comes into position the drawbridges lower onto the ends of the cage and can float up and down with the cage consistent with the rope stretch.

(2) To have supports coming out from the fixed landing on which the cage floor can land. In mining terminology these are known by three different names - Fallers, Props and Keps.

4   THE SHAFT DIAMETER, DEPTH OF SHAFT AND THE SHAFT GUIDE SYSTEM

4.1 The Shaft Diameter

In the last few years a number of new coalmine shafts have been made for British Coal Corporation. These have been about eight metres diameter and this would appear to be suitable for the large radioactive waste package.

The winding system which gives the largest floor area for the cage is a single cage and counterweight system in the mineshaft.

The shaft plan area has also to be large enough to carry electric cables, water mains, compressed air mains, and of course the conveyances rigid guiding system comprising vertical steel guides.

The above information is deduced by drawing a plan view of the vault transfer vehicle and then using trial and error methods for:

1   Drawing the cage and counterweight system around the vehicle

2   Constructing the minimum diameter shaft circle around the cage and counterweight system.

3   Checking that all the other necessary equipment can be fitted in such as pipes, electrical cables etc.

4   Checking that the ventilation air speed is at an acceptable level.

5   The geological and hydrogeological conditions at the site will enable calculations to be done to indicate whether or not the ideal shaft diameter is practical to construct.

4.2 Shaft Depth

In British and most European mines, the average rise in temperature is one degree centigrade for every 30 metres of depth, but the gradient varies in different parts of the world. In coal mines near Manchester, for example, the temperature at a depth of 1220 metres has risen to 45 degrees centigrade, which is already too hot for comfortable working. But in South Africa a temperature not higher than 38 degrees centigrade is usual in the 2400 metre deep gold mines near Johannesburg.

The worlds deepest mine is the Western Deep, near Johannesburg South Africa. It goes down more than 3300 metres.

At 2400 metres in Britain the temperature would be 80 degrees centigrade. The temperature gradient in South Africa is therefore only about half that of Europe, and extension of mining development to 3700 metres is contemplated. In North America the temperature gradient is about three quarters that of Europe. This variation may be due to slight differences in thermal conductivity of the upper layers of the shell; temperature is certainly higher under layers of unconsolidated sediment than in compact siliceous crystalline rocks. The distribution of radio-active minerals may also be important.

The required depth for the shaft under consideration is expected to be between 500 and 1000 metres depending on the geological and hydrogeological conditions of the site selected for the repository.

4.3 Shaft Shape

Mineshafts of rectangular and oval shape have been constructed but the standard form for the majority of shafts is circular.

4.4 The Shaft Guide System

The size of conveyance and counterweight compared with the size of practical shaft diameter construction makes rigid guides the automatic choice from the point of view of running clearance.

4.5 Fixed Guides (Rigid)

It is modern practice to manufacture fixed shaft guides from hollow section, steel tube typically 200 x 200 x 10 millimetres. Guide rollers can be used at high winding speeds. Typically 20 metres per second.

4.6 Rope Guides (Flexible)

Absolute smoothness of conveyance running can be achieved only through the use of rope guides. However, a greater clearance is required around the conveyance on rope guides than on rigid guides.

Generally minimum shaft clearances for rope guides can be taken as:		
DEPTH OF SHAFT	CLEARANCE	
	BETWEEN CONVEYANCES	CORNER OF CONVEYANCE TO SHAFT WALL
Down to 500 m Below 500 m	306 mm 610 - 915 mm	229 - 305 mm 457 - 610 mm
Clearances for fixed guides are generally:		
	305 - 457 mm	152 - 229 mm

5   THE WINDING SYSTEM - CONFIGURATION OF CONVEYANCES AND ROPES IN THE VERTICAL SHAFT

It has already been decided that a multi-rope tower mounted friction winder is required for this application. This would have eight winding ropes and six balance ropes. Also it has been decided that the best system of conveyances in the shaft for the large package to be handled is a single cage and counterweight system on rigid guides.

When the shaft plan is drawn out to scale the position and plan area of the cage and counterweight are as shown in Fig VI - the position of the winding ropes and the rigid guides are also indicated.

6   THE SHAFT TOP AND INNER TOWER EQUIPMENT

6.1 Shaft Top

The area around the shaft top is designed as a loading and reception area for the cage. To protect personnel from falling into the shaft, all the four sides of the shaft are covered with fencing or gates.

Rail tracks are provided on the landing floor and at the edge of the shaft the retractable hinged fallers are positioned.

There are two rail tracks, the one in line with the cage is for full vault transfer vehicles.

6.2 Inner Tower

Inner towers may be designed to be suspended from the main winder tower or they are designed to be free standing on concrete foundations at ground level.

The inner tower is a steel structure comprising vertical columns, horizontal beams and diagonal bracings.

This steel structure carries the following equipment:

(a) Bumping beams to withstand winding ropes breaking load.

(b) Overwind safety catchgear.

(c) Shaft arrester gear, to retard the winding system from full speed and bring it to rest at a rate of deceleration not exceeding 'g' (9.81 m/sec/sec) (32.2 ft/sec/sec).

(d) Access stairway from ground to bumping beams level.

(e) Platforms, fencing, handrails at convenient levels to give access to all equipment within the tower.

(f) Receiving and positioning guides.

(f) Gates and fencing at ground level and at man landing levels.

(h) Rope installation pulleys.

(i) Cable installation pulleys.

(j) Conveyance installation and changing equipment.

(k) Electrical switches and lighting.

6.3 Shaft Bottom

The area around the shaft is designed as a reception area for the cage as it reaches the fixed landing.

The openings opposite the ends of the cage are fitted with gates and the other sides of the shaft are fenced off to make then safe for personnel approaching the shaft side.

An overhead canopy is provided where the landing projects out into the eight metre diameter shaft area to reach the ends of the cage. This canopy is to protect personnel, approaching or leaving the cage, from debris which may fall down the shaft.

Rail tracks are provided on the landing floor and at the edge of the shaft the hinged fallers are positioned.

The underground arched roadways are eight metres wide and two railtracks are laid along this roadway which leads from the shaft to the vaults. The railtrack which is directly in line with the cage is for the full vehicles while the other railtrack is for the returning empty vehicles.

This railtrack passes the shaft and at the far side a traverser is fitted, a device which can move the vehicle from one track to the other.

In simple terms it is steel platforms level with the roadway floor and has the two rail tracks fitted to it. The platform is longer than the vehicle. When the vehicle is on the Traverser it can be moved sideways until the vehicle is in-line with the second track.

7   THE SHAFT BOTTOM SUMP EQUIPMENT

7.1 Sump Steelwork

In friction winding installations the equipment contained in the sump is similar to that in the surface inner, tower namely:

(a) Access ladders from roadway level to shaft bottom.

(b) Platforms, fencing, handrails at convenient levels.

(c) Receiving and positioning guides.

(d) Gates and fencing at fixed landing level.

(e) Balance rope boxes.

(f) Chairing equipment for conveyance rope changing.

The traditional manner of supporting the steel frames in the sump is by excavating pockets in the mineshaft walls and concreting stub girders into these to support the frames. An alternative method of support for the steel frames is to bolt brackets onto the mineshaft wall.

To avoid making wall pockets or drilling for brackets it is possible to build the sump steelwork as a tower structure such that the bases of the columns are bolted to the shaft bottom. Horizontal restraint is provided at each frame level by multipacks/shims secured between the frame and the mineshaft wall.

8   THE SHAFT CONVEYANCES
CAGE AND COUNTERWEIGHT AND SUSPENSION GEAR

8.1 Cage

The cage is basically a steel box which carries the load between the top and bottom of the shaft in a similar manner to a lift car for carrying passengers in a building.

In mining the minimum factor of safety for the steel framework of the cage is seven to one. The plan length is seven metres and the plan width 3.25 metres. The height is eight metres for the load weighing 85 tonnes.

The steel framework which forms the roof of the cage is designed to take the pin connections from the eight strands of winding rope suspension gear.

The floor of the cage comprises a steel framework fully plated over and carrying the rails for the vault transfer vehicle.

Below the floor of the cage the steel framework is designed to take the pin connections from the six strands of balance rope suspension gear.

8.2 Counterweight

The counterweight comprises a steel framework designed to carry a number of identical metal weights which act as the ballast.

The steel framework is designed to a factor of safety of seven to one. The top of the frame is designed to take the pin connections from eight strands of winding rope suspension gear and the bottom of the frame is designed to take the pin connections from the six strands of balance rope suspension gear.

8.3 Suspension Gear

This is equipment which is used to connect the ropes to the shaft conveyances.

The Winding rope suspension gear comprises:

i   A capel, which is a device secured to the rope at one end and has a socket and pin connection at the other end.

ii   A number of links and pins.
These connect together and form a flexible chain. The links are designed to give maximum flexibility in two planes and vary in length to give a capability for coarse and fine length adjustment. This is to assist with rope shortening and equalising the tensions in the eight ropes.

One link in each strand is fitted with a load cell which is used to compare the load in each rope and also show the total load in the system ie, cage, load and balance ropes.

The balance rope suspension gear is similar to the winding rope suspension gear in that it connects the ropes to the underside of the conveyance and gives flexibility in two planes.

A swivel device is incorporated in each strand which allows the ropes to turn backwards or forwards so that they do not get twisted.

9   ELECTRICAL CONTROL SYSTEM

Due to the nature of the plant the electrical control equipment, in the form of dual programmable logic controllers (PLC's), would be housed in a safe, hazard free, controlled environment at the surface of the shaft.

The plant transducers and output devices (solenoids etc) which are located in potentially hazardous areas would be hard wired to the host controller (PLC). To provide the maximum integrity both the transducers and output devices would be duplicated.

9.1 Advantages and Facilities of PLC Control

When compared with conventional relay control systems the advantages of the proposed systems would be:

MODULARITY - Large and small PLC's can be constructed from combinations of the same standard building blocks. The customisation to suit a particular control task comes in the form of an easily alterable programme which may be written and modified by the end user.

FLEXIBILITY - Is inherent to modular construction, whereby input/output (i/o) plug-in modules can be added or removed and the revised control software programme amended to suit. Redeployment of the PLC onto other differing control applications is therefore possible.

CONTROL COMPLEXITY - A PLC imposes virtually no limit on the logic control scheme in terms of number and complexity of boolean combinations. The enhanced level of complexity allows advances features to be incorporated in the control scheme such as plant monitoring and optimisation.

DISPLAY - The current status of (I/O) channels may be observed from indicators situated on the associated PLC module. The system status may be ascertained from a numeric indicator situated on the central processing unit (CPU) module.

DIAGNOSTICS AND PROGNOSTICS - Spare CPU time can be utilised to give powerful self-diagnostic features, offering valuable guidance to artisans in the event of system and plant faults. There is also the possibility of prognostic features in the form of routing condition monitoring to pre-empt faults and report any intermittent plant or system malfunctions.

UPGRADES - The rate of change in improvements to micro-electronics is far higher than that of other control devices. With a structured, modular design approach, the original module format may be used to retain compatibility with earlier designs. This is important where it is necessary to incorporate the latest technology.

PHYSICAL SIZE - The PLC occupies less space than the equivalent relay system, however certain users may insist on traditional link type terminals for plant I/O cables, thus tending to minimise the reduction in comparable cubicle sizes.

RELIABILITY AND MAINTENANCE - High standards of PLC reliability can be achieved and the absence of moving parts along with self check features minimises maintenance in the field.

COMMUNICATIONS - Serial multiplexed communication links replace the traditional multicore cables to transmit information. These may be employed over long distances.

RESPONSE TIME - A fast response time can be realised even for system calling for lengthy plant control software programmes.

USER FRIENDLY - PLC's are generally easy to use and configure, and are provided with relatively simple means of programming user relay ladder logic.

COMMISSIONING - PLC's have facilities which permit changes to the plant control programme in situ.

9.2 Safety Considerations

Safety is a most important requirement for plant control systems, the various aspects of which are described.

9.3 Operational Safety

This encompasses the complete range of plant operation and control performed by the system, together with any fault conditions that may occur. If no danger to personnel or equipment can arise, then the system is operationally safe.

Within these considerations the output devices being controlled must also revert to a safe condition in the event of a fault of PLC supply failure.

9.4 Fail Safety

A fail safe PLC operating system should ensure that under error conditions operation will terminate in a predetermined safe state, provided it is not subject to a common mode failure. terminate in a predetermined safe state, provided it is not subject to a common mode failure.

9.5 Operating Principles

The PLC unit spare time in which it may execute other tasks such as checking the integrity of internal registers and verifying the operation of input and output interfaces along with self regulation. Should a major error be detected, the CPU can force outputs to a predetermined safe state, independent of current plant circumstance. However, facilities of this kind are not often provided.

9.6 Emergency Facilities

United Kingdom mining applications usually call for hard wired emergency stops which may be provided by a relay safety circuit. Operation of an emergency stop strips the safety relay, which inhibits the PLC and possibly removes the output channel load supply. The relay may also be tripped by the PLC in the event of a fault being detected internally.

10   GLOSSARY OF TERMS USED IN MINING PRACTICE

10.1 Automatic Contrivance

The apparatus provided to prevent overwinding or overspeed of a manriding conveyance or counterweight and which is designed:

(a)   To prevent the descending conveyance from being landed at the lowest entrance or at the bottom of the shaft at an excessive speed; and

(b)   to control movement of the ascending conveyance to prevent danger to persons therein; and

(c)   to control the speed of the conveyance or counterweight and prevent travel beyond predetermined end of wind limits.

10.2 Balance Rope

A rope suspended from and linking the underside of the conveyances, or conveyance and counterweight, to reduce the effect of the out-of-balance weight in the shaft.

10.3 Buntons

A horizontal steel beam within the shaft normally supporting rigid guides, pipes, cables etc.

10.4 Cage

A cage is a horizontal platform with sides and a roof to carry personnel and equipment up and down the shaft. At each end of the platform a safety gate is fitted.

10.5 Capping/Capel

The complete attachment fitted at the end of a rope, by means of which the rope is coupled to any suspension gear or conveyance or counterweight.

10.6 Downcast Shaft

To provide a ventilation system for underground tunnels there are usually at least two shafts. One of these is called the downcast shaft and the ventilation air is drawn down this shaft by the power exerted by the ventilation fan.

10.7 Factor of Safety

The ratio between the ultimate tensile strength of a component and the maximum static load to which it may be subjected.

10.8 Fallers, Props, Keps

Retractable supports in a shaft on which a conveyance may rest.

10.9 Friction Winding System

A winding installation in which a rope or ropes are attached to a conveyance or counterweight at each end and in which movement of the conveyance or counterweight is produced by friction between the rope or ropes and the treads of a driven drum or sheave of a winding engine.

10.10 Guides

An arrangement of girders, rails, timbers or ropes, in a shaft to restrict lateral movement of conveyances or counterweight.

10.11 Overwind

Unintentional overtravel of a conveyance or counterweight beyond the normal limits of winding.

10.12 Overwind Safety Catches

Catches or equivalent devices provided in headframes to prevent conveyances from falling back an excessive distance after a severe overwind, independent of the suspension gear.

10.13 Suspension Gear

Apparatus including any detaching hook for attaching to the winding rope a conveyance or counterweight.

10.14 Upcast Shaft

To provide a ventilation system for underground tunnels there are usually at least two shafts. One of these is called the upcast shaft and the

10.15 Vault Transfer Vehicle

This is a purpose made locomotive flat top for carrying the package. In is self powered with integral braking system and is operated by remote control.

Fig I   PHOTOGRAPH OF TYPICAL WINDER TOWER
AND INNER TOWER

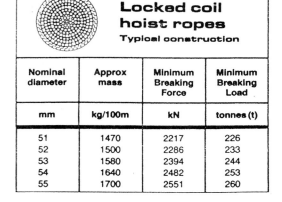

Locked coil hoist ropes
Typical construction

Nominal diameter	Approx mass	Minimum Breaking Force	Minimum Breaking Load
mm	kg/100m	kN	tonnes (t)
51	1470	2217	226
52	1500	2286	233
53	1580	2394	244
54	1640	2482	253
55	1700	2551	260

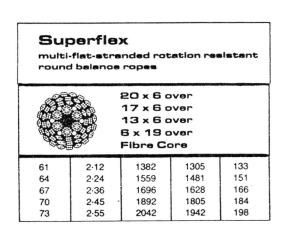

Superflex
multi-flat-stranded rotation resistant round balance ropes

20 x 6 over
17 x 6 over
13 x 6 over
6 x 19 over
Fibre Core

61	2·12	1382	1305	133
64	2·24	1559	1481	151
67	2·36	1696	1628	166
70	2·45	1892	1805	184
73	2·55	2042	1942	198

Fig III   WINDING ROPES AND BALANCE ROPES

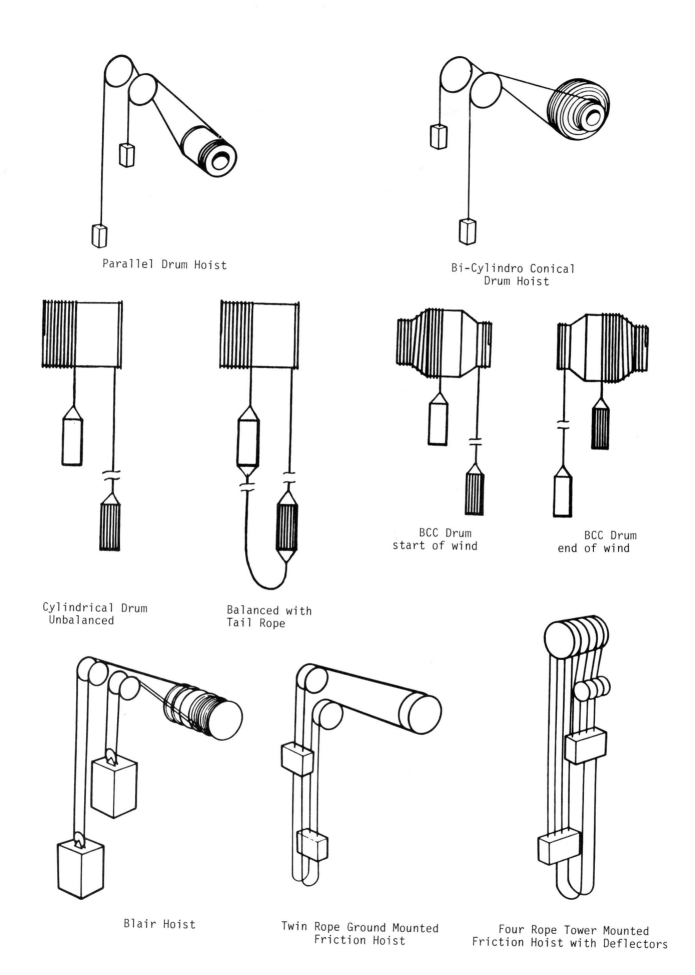

Parallel Drum Hoist

Bi-Cylindro Conical
Drum Hoist

Cylindrical Drum
Unbalanced

Balanced with
Tail Rope

BCC Drum
start of wind

BCC Drum
end of wind

Blair Hoist

Twin Rope Ground Mounted
Friction Hoist

Four Rope Tower Mounted
Friction Hoist with Deflectors

Fig II   VARIOUS WINDING SYSTEMS SHOWING THE CONFIGURATION OF CONVEYANCES
AND ROPES IN THE VERTICAL SHAFT

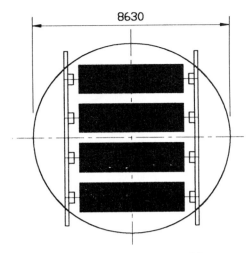

MALTBY NUMBER 3 SHAFT
4 x 25 TONNE

ASFORDBY MINE
2 x 26 TONNE

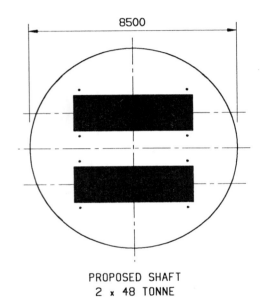

PROPOSED SHAFT
2 x 48 TONNE

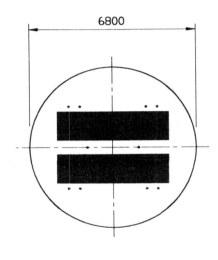

HARWORTH EXISTING NO.1
SHAFT 2 x 27.5 TONNE

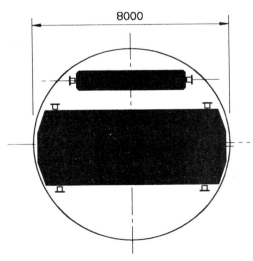

WASTE SHAFT LAYOUT (1 x 85 TONNE)
SHOWING CONVEYANCE AND GUIDE SYSTEM

Fig IV   SHAFT PLANS SHOWING TYPICAL GUIDE
            AND CONVEYANCE POSITIONS

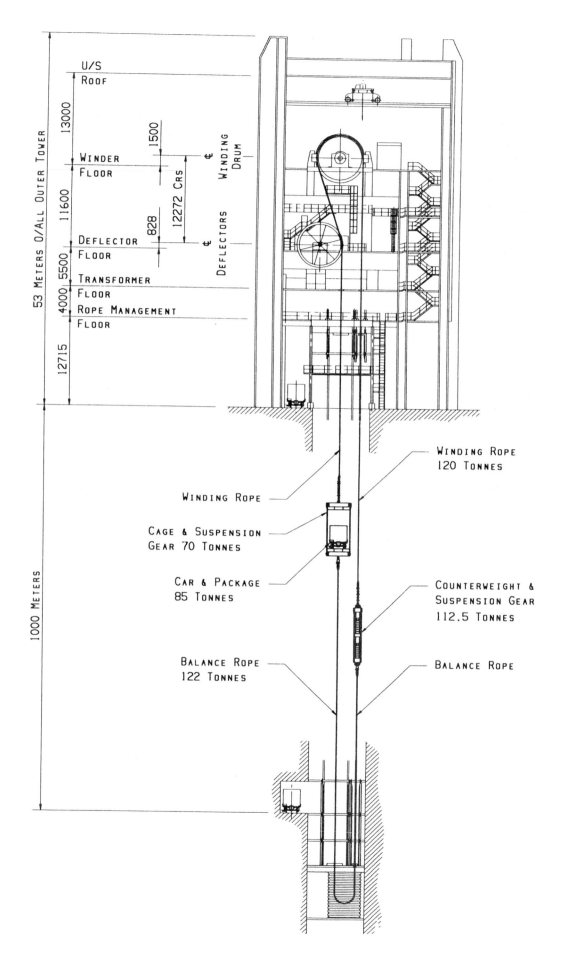

Fig V   LONGITUDINAL SHAFT SECTION SHOWING WINDING SYSTEM

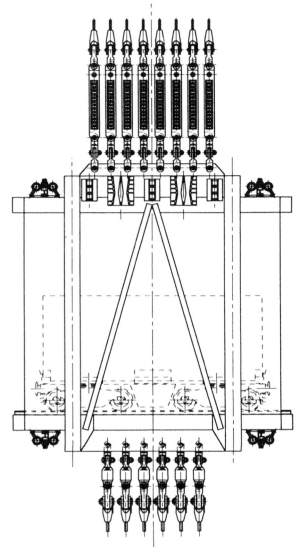

ELEVATION OF 70 TONNE CAGE & SUSPENSION GEAR
WITH 85 TONNE WASTE CAR

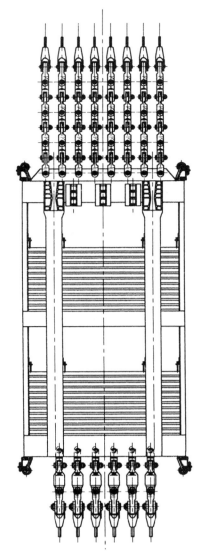

ELEVATION OF 112.5 TONNE COUNTERWEIGHT
& SUSPENSION GEAR

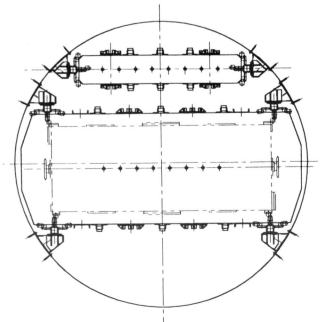

PLAN OF 8.0 M DIAMETER WASTE SHAFT SHOWING
RELATIVE CAGE & COUNTERWEIGHT POSITIONS

Fig VI   CAGE, COUNTERWEIGHT AND
             SUSPENSION GEAR

162

C431/043

Transfer systems in an underground repository

H P BERG and D EHRLICH
Bundesamt für Strahlenschutz (Federal Office for Protection Against Radiation),
Salzgitter, Germany

SYNOPSIS In addition to logistic problem definitions taking into account the waste types of the wastes to be disposed of and the mining conditions, transport and handling of radioactive wastes in a repository particularly require the keeping of safety technological marginal conditions mainly resulting from the accident analyses carried out. The realization of these safety technological aspects is described taking the planned Konrad repository as an example.

1 INTRODUCTION

In the Federal Republic of Germany, two repositories for solid radioactive wastes are planned, both in deep geological formations. The Konrad mine is a former iron ore mine where it is intended to dispose of radioactive wastes with negligible thermal influence upon the host rock. The Gorleben site is at present under consideration for suitability as a repository for all kinds of radioactive wastes including heat-generating wastes and for direct disposal of fuel elements. The planned Gorleben repository shall be built just for the task of final disposal.

Hence, the planned transfer systems in these two repositories have to be different taking into account the structure of the mining and of the waste packages. In particular, the emplacement of packagings containing fuel elements with a mass of about 65 tons in the Gorleben repository necessarily leads to a concept of a rail-guided transport system in the underground facility.

The role of the Morsleben mine, which is located in the area of the New Federal States and in which radioactive wastes from the former GDR were emplaced until the end of 1990, still is uncertain, because additional safety measures will have to be taken at any rate. With regard to the underground emplacement logistics, the components determined for the planned Konrad repository would be orientated to. Thus, the explanations in the following chapters are restricted to the ante-conceptional phase of the Gorleben plannings and the detailed works for the planned Konrad repository. More details about the Konrad repository, in particular with respect to safety aspects of the operational and postoperational phase, are given in (1) and (2).

2 CONCEPTUAL CONSIDERATIONS FOR THE PLANNED GORLEBEN REPOSITORY

While, due to the advanced planning state, there already exist very detailed requirements on the design of the planned Konrad repository (see chapter 3) up to the construction of prototypes of single components of the emplacement process, only first reflections on the planned Gorleben repository have been made. This is, on the one hand, caused by the fact that the operation of the Gorleben repository is not planned to be started before the year 2008 and the site investigation is still being carried out at present, on the other hand, by the fact that a number of emplacement techniques within the framework of r&d-works will have to be developed and proved.

While, in analogy to the planned Konrad repository, the stacking technique is to be applied for the emplacement of low heat-generating radioactive wastes, two other emplacement techniques are planned for the heat-generating wastes according to their conditioning.

The lowering technique into 300-m and even into 600-m deep boreholes is planned for vitrified wastes packaged in canisters as well as for cemented wastes packaged in drums; tests of the handling facilities at present being developed are planned to be carried out in the Asse mine on a scale of 1:1.

Emplacement in galleries as a concept is planned for the emplacement of fuel elements in Pollux containers (direct emplacement).

Also being proved in an r&d-program are the technical realizability of the underground emplacement system for the emplacement in galleries including the essential components with regard to engineering and mining as well as the safety of the transport and emplacement operation with regard to radiological protection of the staff.

As in the Konrad mine, platform trucks are planned to be used for internal transportation, which are supposed to transport all containers - i. e. the ones weighing 65 tons too - with the wastes to be disposed of directly. Such platform trucks already exist, however, up to now, they are only used above ground, for instance for the transport of coal.

In comparison, mining locomotives, also used for the transport of great masses, form part of the state of the art. For reasons of fire protection, however, a battery-driven locomotive is considered reasonable.

The components platform truck and emplacement facilities with the help of which heavy containers are unloaded from the platform truck, are constructed and tested within the framework of the above-mentioned r&d program.

3 TRANSFER SYSTEMS IN THE PLAN-NED KONRAD REPOSITORY

The former Konrad iron ore mine is located in the southeast of Lower Saxony between Braunschweig and Salzgitter-Lebenstedt and is operated by two shafts. When being operated as repository, the Konrad 1 downcast shaft is to serve for man-riding, debris haulage, as well as for the transport of materials. In the Konrad 2 upcast shaft, haulage of waste packages underground is carried out.

All surface installations belonging to the Konrad 2 shaft, i. e. all buildings including the hoisting plant will newly be constructed for the operation of the repository.

Because of the special licensing procedure for a repository for radioactive wastes, the Federal Office for Radiation Protection being applicant here, all phases of the project including operation must already be assessed and authorized on the basis of the planning documents, so that the documents to be submitted are already very detailed at an early stage. Some of these plannings are described in the following.

3.1 Sequence of emplacement operations

The operations in the planned Konrad repository starting with the delivery of waste packages (i. e. of the processed and packaged radioactive wastes) to the shaft area up to their stacking in a depth between 800 m and 1300 m, are summarized in figure 1. Within the framework of detailed accident analyses,

the events which might lead to a release of radionuclides from the waste packages have been identified.

By radiological calculations it has been shown for the incidents "dropping of a transport unit from a subsurface waste transfer station from a height of 3 m, dropping of a waste package from a height of 5 m when being stacked in a subsurface emplacement room, as well as subsurface fire of a means of transport, that the incident planning values of the Radiation Protection Ordinance are fallen below.

The transport units (one container or up to two cylindrical waste packages on a pool pallet) are delivered by wagon or by truck.

The transportation of the wagons is carried out by German Federal Railways up to the Beddingen delivery station. Here, the wagons are taken on by "Verkehrsbetriebe Peine-Salzgitter GmbH" and transferred to the Konrad 2 shaft area where the transport units are transported below ground. From the north, the Konrad 2 mine can be reached by truck via an access road belonging to the mine.

After access is given, the trucks drive to the planned truck parking lots or directly in front of the truck drying plant.

After access to the mine area is given, the wagons are hauled by a company switch engine to a buffer rail where they are parked and, according to wagon type, uncoupled in units of two or three wagons and pushed to the drying plant.

If necessary, water, snow and ice are removed from the wagons or trucks in the drying plant.

The transloading of the transport units from truck or wagon to platform trucks is carried out with a bridge crane in the transfer hall.

The platform trucks loaded with transport units are driven with the track haulage installation to the incoming controlling device operated by personnel responsible for radiological protection. Here, for instance, surface contamination is checked and the waste package specific local dose rate is measured.

When the staff responsible for radiological protection has released the transport unit, the platform truck is as a rule driven in front of the safety gate in the shaft hall.

For transport units, however, there is also a buffering possibility in the plant. A battery-driven side stacking vehicle is used for parking and removing transport units from the buffer hall.

With the hoisting plant, the platform truck is driven to the 850-m level.

Before the unloading at the shaft bottom is started, the transport vehicle drives in reverse movement from the emplacement transport gallery to the separator in loading position.

The transport unit is dumped from the platform truck with a rail-guided portal lift truck which drives over the transport vehicle and puts the transport unit down on the transport vehicle's load area.

With the transport units, the transport vehicle shuttles through the emplacement transport gallery from the shaft bottom to the unloading room of the emplacement room and back to the shaft bottom.

Inside the unloading room, the stacking vehicle takes over the pool pallet or the container from the transport vehicle.

By retreating, the emplacement rooms are filled with waste packages, the cylindrical waste packages horizontally and the containers standing on their bases.

The remaining voids between the stacked waste packages and the stope or the roof are stowed with pumping stowage.

3.2 Components of the underground transfer system

When choosing the emplacement system components, different aspects had to be taken into consideration, like minimization of the number of unloading operations, the possibility to safely immobilize the transport units on the transport vehicles, use of uniform slings with different handling methods above and below ground, as well as application of systems tested in mining with which incidents are not likely to occur. On the other hand, surfaces with decontaminational features are desirable with regard to radiological protection.

Spreader technique is mainly to be used for unloading. To make this possible, a list of standardized containers has been drawn up in co-operation with the waste producers who are obliged to deliver their radioactive wastes to a repository (table 1). Here, at the maximum two cylindrical waste packages per pallet are delivered and transported to the emplacement room.

Essential criterion for the choice of the platform truck as internal transport system was, among other things, that the transport units can be immobilized on the cage during transportation of the waste packages through the shaft below ground.

Below ground, the following machines and vehicles are used for the process of emplacement:

- hoisting plant,
- portal lift truck,
- transport vehicle,
- traffic control facilities, and
- stacking vehicle.

The portal lift truck, transport vehicle and stacking vehicle which were already constructed as prototypes and which are at present being tested above ground, are described in the following.

When planning and constructing these means of unloading and transport, it was, if possible, fallen back upon systems already available and tested in mining or at least upon single components. In spite of this, a new construction was necessary for the special needs of the planned Konrad repository, so that for both vehicles a design licence was applied for at the responsible Higher Mining Office.

In the shaft bottom, a rail-guided portal lift truck is used for unloading the transport units from the platform truck onto the transport vehicle and the empty pool pallets from the transport vehicle onto the platform truck.

The portal lift truck consists of triple portal, rail landing-gear, cable hosting gear, lifting device and load lifting device.

The load lifting device for transport units consists of 2 sliding beams to which a centring frame with 4 adjustable twist locks is attached which is movable by chain links. The twist locks can electromechanically be adjusted to the different percussion points of the transport units. The sliding beams fixed in the lifting frame are adjustable diagonally to the driving direction. In connection with the rail landing gear, this enables an exact positioning of the centring frame for lifting and putting down the transport units.

With the help of a lift limit stop device, the height of lift between shaft bottom and bottom edge of the transport vehicle is limited to a maximum of 1.90 m. Between platform truck and transport vehicle there is a separator, so that in case of a drop the transport unit will drop on the separator and not on the level of shaft bottom. This reduces the dumping height (bottom edge of transport unit - top edge of separator) by approximately 1 m. This dumping height is assumed in the radiological calculations.

For the transportation of the waste packages from the shaft bottom to the emplacement room transport vehicles are used. Spiral chutes - with correspondingly narrow radius of curve - and

Table 1 List of standardized containers

| No. | Designation | External dimensions | | | Gross volume |
		Length/ diameter mm	Width mm	Height mm	m³
01.	Cylindrical concrete packaging type I	⌀ 1060	–	1370 [1])	1,2
02.	Cylindrical concrete packaging type II	⌀ 1060	–	1510 [2])	1,3
03.	Cylindrical cast iron packaging type I	⌀ 900	–	1150	0,7
04.	Cylindrical cast iron packaging type II	⌀ 1060	–	1500 [3])	1,3
05.	Cylindrical cast iron packaging type III	⌀ 1000	–	1240	1,0
06.	Container type I	1600	1700	1450 [4])	3,9
07.	Container type II	1600	1700	1700	4,6
08.	Container type III	3000	1700	1700	8,7
09.	Container type IV	3000	1700	1450 [4])	7,4
10.	Container type V	3200	2000	1700	10,9
11.	Container type VI	1600	2000	1700	5,4

[1]) Height: 1370 mm + 90 mm lifting lug = 1460 mm
[2]) Height: 1510 mm + 90 mm lifting lug = 1600 mm
[3]) Height: 1370 mm for KfK type
[4]) Stacking height: 1400 mm for KfK type

Container materials are e. g. sheet steel, reinforced concrete or cast iron.

ramps with a slope of up to 12 % have to be driven on. This cannot be done with rail-guided vehicles.

The transport vehicle therefore is a four-wheel driven railless vehicle which consists of a motor part and a load part. Figure 2 shows the prototype of the transport vehicle being tested above ground. The construction enables the necessary maneuvering capability and curve negotiating characteristic on the emplacement galleries from the shaft bottom to the emplacement rooms. The transport vehicle is driven by a Diesel engine. Due to the technical construction, its travelling speed is limited to a maximum of 4 m/s.

The load part is formed as loading space for taking up the different transport units and equipped with a continuously variable locking device for centring and securing the transport units on the loading space.

The thermal loads of the transport vehicle are limited to about 700 l oil and approximately 1.700 kg rubber. This guarantees that the temperature-time-curve assumed as a model - but covering - for a fire below ground is not exceeded. Fire of a vehicle below ground is treated as a radiologically representative occurrence in the safety analyses, though the vehicles are provided with board fire extinguishers which are equipped with thermometer probes and fire extinguishing nozzles for motor, transformer and gear unit.

The transportation of the transport units in the emplacement rooms as well as their emplacement are carried out by stacking vehicles.

The stacking vehicle is a four-wheel driven railless vehicle and consists of a motor part and a load part which are connected with a buckling link. The stacking vehicle is driven by a Diesel motor.

For picking up the transport units, the following, easily changeable equipment is available:

- fork-tines to handle pool pallets and cylindrical waste packages,
- side-spreaders with spreader hooks which are continuously adjustable to the dimensions of all containers.

From the driver's compartment, the fork-tines and spreader hooks can be adjusted to the different reference and percussion points of the waste packages with the help of a stored programmable steering mechanism.

The stacking vehicle's height of lift is limited in such a way that the - with regard to its outer dimensions - lowest waste package (cast-iron container type I with a diameter of approx. 0.9 m) can dump on the bottom of the emplacement room from a maximum height of 5 m.

With 2.78 m/s, the driving velocity of the stacking vehicle is even below the one of the transport vehicle. Because of this, possible mechanical influences on the transported waste package are reduced when driving against the stope.

The technical equipment of the stacking vehicle with regard to fire protection is comparable with the one of the transport vehicle.

166

© IMechE 1991 C431/043

This chapter describes the tranfer system in the underground part of the repository. Comparable requirements have been derived for the means of transportation in the above ground facility (transfer hall and buffer hall, cf. figure 1). They consist of special fire protection measures, in particular with respect to the delivery trucks, and limitations of the driving velocity of the lateral stacker truck as well as limitations of the height of lift of the bridge crane and of the lateral stacker truck.

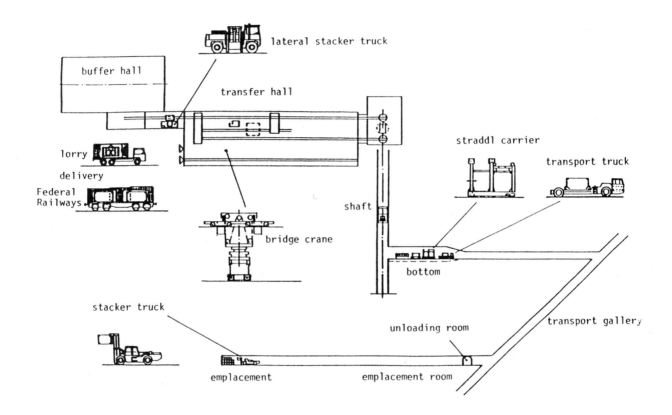

Fig 1   Planned emplacement operations in the Konrad repository

Fig 2   Prototype of the transport vehicle

C431/043 © IMechE 1991

167

The goal of the planning of the transport and handling system was to guarantee a trouble-free and safe progress. This safety has been reached by the corresponding layout of the individual components, the suitable choice of the corresponding means of transport, and the easy coordination of the different components, so that the demands made by the planned Konrad repository from a logistic and safety technological viewpoint are kept. This includes the design of the hoisting plant described in the following chapter.

3.3 Design of the hoisting plant

Extensive practical experience in mines has been gained in transport operation in the shaft. This experience is reflected in the technical requirements for hoisting plants and inclined conveyors and in the additional requirements of the mining authorities for the design, operation and supervision of such plants. In this way a high degree of safety is generally guaranteed in the operation of hoisting plants in shafts.

In addition, essential safety elements which serve to prevent damage are part of the design of the Konrad 2 hoisting plant. Examples are:

- A multirope hoisting plant has been provided in the Konrad 2 shaft.
- The mass of each waste package will be controlled.
- The transport unit and the platform truck are adequately fixed during transport in the shaft.
- The brakes of the winding engine and the stopping distances at the end of the guide rails will be specially designed.
- The speed control is checked continuously and at certain points.
- Loading and locking devices for the cage, which are included in the winding engine's safety circuit, will be used at the pit-head bank and at the bottom.

Among the design basis incidents which may occur in the area of the hoisting plant is, for example, the dropping of a platform truck during loading of the cage. The precautionary measures taken in order to avoid the dropping of waste packages are described below.

The shaft is located symmetrically between the approach track and the return track. Each of the tracks, running in parallel, ends at a turntable equipped with telescopic pulling and pushing devices. The track haulage device of the transfer facility positions the platform truck on the approach track in a truck stop located in front of the turntable. The pulling device pulls the truck on to the turntable which, for this purpose, has been turned into the direction of the approach track.

The loading operation from the turntable to the cage starts when:

- the brakes of the winding engine have been locked;
- the cage floor rests on the plugs;
- the turntable has moved to the direction of the cage;
- the empty platform truck has been taken up by the turntable located in the return track.

Having been pushed up, the platform truck loaded with a waste package is positioned in recesses in the rails. The truck is fixed to the cage floor by retaining bolts which are operated automatically when the cage floor is lowered.

The design described above for the loading and locking devices guarantees that the dropping of waste packages during loading of the cage and physical damage to the waste packages due to the dropping of a platform truck into the shaft are avoided.

4 VENTILATION CONDITIONS IN THE PLANNED KONRAD REPOSITORY

With regard to location and ventilation, the emplacement area - which is also control area - is separated from those areas in which new openings are driven, i. e. in which conventional mining is carried on. On the one hand, this is reached by the fact that temporally running ahead the necessary openings are gradually driven, on the other hand a parallel ventilation of the respective emplacement area and the respective conventional area occurs by so-called own ventilation districts. The conduction of potentially contaminated return air from emplacement rooms to the uptake shaft occurs through special galleries in which there are no places of work permanently occupied.

Furthermore, there are planned separate transport paths for the waste packages to be emplaced and the debris from the excavation of the openings which partly is to be hauled. The waste packages are transported through the uptake shaft below ground and the necessary debris haulage as well as the transportation of personnel is carried out through the intake shaft. The stowage material is prepared at the boundary between the conventional and the emplacement area and - separately from the emplacement transportations from the point of view of time - transported to the waste packages to be stowed.

The installations necessary for the ventilation of the mine are designed in such a way that air supply of all relevant mine workings as well as dilution and drawing off of damaging gases caused by Diesel-driven vehicles are guaranteed.

Therefore, the places of operation must be provided with at least 3.4 m^3 fresh air per minute per installed Diesel kW for the vehicles driven in there. Because of the vehicles used, up to 103 m^3/s fresh air are necessary only inside the emplacement area.

In order to provide the mine with a sufficient amount of air requirements at any time, the main fan, which is located above ground at the Konrad 2 shaft, is designed for an air quantity of 260 m^3/s.

Within the framework of the main ventilation, fresh air casts down the Konrad 1 shaft, is distributed in the mine on the different galleries and casts up in the Konrad 2 shaft. Air flow distribution in the mine is carried out by the installation of air locks and air regulators. The task of air locks is to separate two airways from each other and to enable traffic of staff and vehicles between the galleries in spite of this.

With the help of adjustable openings, air regulators reduce the air flow running through to the intended degree by decreasing the roadway cross section.

The main ventilation is designed in such a way that in all operating conditions mine air from the controlled area cannot reach the conventional area.

A suction auxiliary ventilation of the emplacement rooms is carried out during the emplacement operation. The air of the in situ area is exhausted with a fan located in the return air central road through a ventilation borehole and a string of air ducts.

With the help of the planned ventilation and area concept an uncoupling of the different operations in the repository and, together with this, a minimization of unwanted occurrences is gained. In the case of an activity release, only a small part of the mine is contaminated, and only a low number of staff members is concerned. A contamination of the hauled debris is excluded.

The basic ventilation conditions for the planned Konrad repository are illustrated in figure 3.

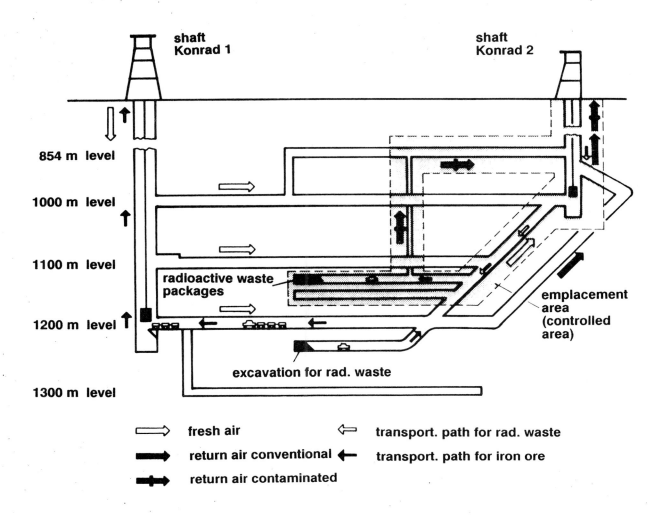

Fig 3 Ventilation conditions

5 CONCLUSIONS

The main goal of the emplacement system of the planned Konrad repository for the disposal of radioactive wastes with negligible heat production is to ensure the safe handling and transport of these wastes. The different components of this emplacement system have been chosen und designed under these conditions taking into account logistic and safety related aspects. In particular, results of the accident analyses have led to requirements on the means of transportation concerning fire protection measures and limitations for example of the driving velocity. Moreover, it is necessary to design the ventilation installations in the underground part of the repository

in such a way that the fresh air supply of all relevant working places of the personnel is guaranteed.

REFERENCES

(1) BERG, H.P., BRENNECKE, P. The Konrad Mine - The Planned German Repository for Radioactive Waste with Negligible Heat Generation. BfS-Bericht ET-6/90, Salzgitter, Juli 1990.

(2) BERG, H.P., EHRLICH, D., ILLI, H., THOMAUSKE, B. Safety Analyses and Derivation of Site-specific Requirements on Radioactive Waste for the Planned German Repository "Konrad". Nuclear Technology, 1987, 79, 92 - 99.